AN IN   NEERS

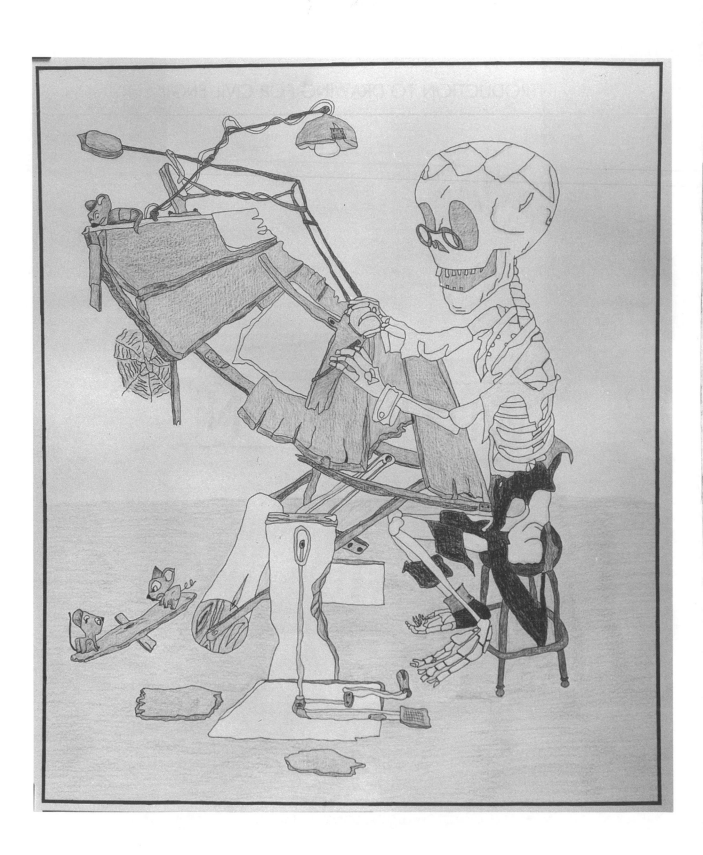

AN INTRODUCTION TO DRAWING FOR CIVIL ENGINEERS

AHMED ELSHEIKH

Department of Civil Engineering
University of Dundee

McGRAW-HILL BOOK COMPANY

London · New York · St Louis · San Francisco · Auckland
Bogota · Caracas · Lisbon · Madrid · Mexico · Milan
Montreal · New Delhi · Panama · Paris · San Juan
São Paulo · Singapore · Sydney · Tokyo · Toronto

Published by
McGRAW-HILL Book Company Europe
Shoppenhangers Road, Maidenhead, Berkshire, SL6 2QL, England
Telephone 01628 23432
Fax 01628 770224

British Library Cataloguing in Publication Data
Elsheikh, Ahmed
    Introduction to Drawing for Civil
    Engineers
    I. Title
    604.2

    ISBN 0-07-709050-0

Library of Congress Cataloging-in-Publication Data
This data is available from the Library of Congress,
Washington DC, USA.

1234 CUP 9765

Typeset by TecSet Ltd, Wallington, Surrey
and printed and bound in Great Britain at the University Press, Cambridge
Printed on permanent paper in compliance with ISO Standard 9706

Dedication

This book is dedicated to my wife Rania and my parents, without whom it would not exist.

Contents

Preface and Acknowledgements

In the technical world of engineering, engineers, designers, draughtsmen and technicians work together as a team to build the many projects that are essential for our modern life. Cooperation and communication of ideas are necessary for this task. While oral and written languages are important, graphical means are also essential for proper communication in an engineering atmosphere.

This book has been designed to introduce the student, who is about to enter the field of engineering, to the science and art of graphical communication. The book is based on the latest standards of drawing conventions accepted worldwide and therefore can serve as a reference book in design offices.

The book is divided into two main parts; the first is on the basics of drawing and general drawing techniques. The second part on the civil engineering applications of drawing builds on the background laid in the first part and introduces the reader to the techniques and conventions adopted in the various fields of civil engineering.

Throughout the book, the reader is presented with self-explanatory examples and plenty of exercises. It is recommended that the readers should try as many of these exercises as possible, and discuss their solutions with their advisers of studies. It is hoped that students following the book will acquit themselves in the essential language of drawing and will find the book interesting.

I am deeply indebted to the many people who have helped during the preparation of this book, particularly Mr Ian Imlach, Mr Alan Davidson, Mrs Mary Whitehorn and Dr Nutan Subedi. I am grateful for their wholehearted support and encouragement throughout the year taken to prepare this book.

Abbreviations and Symbols

The important abbreviations and symbols used throughout this book which are common in civil engineering projects are listed in the following:

aggregates	agg
approximately	approx
average	av
breadth	B or b
British Standard	BS
British Standards Insitution	BSI
centre line	CL
centre to centre	c/c, @
column	col
concrete	conc
dead load	DL
diameter	Φ, D, d or dia
dimensions	DIMS
drawing	drg
elevation	Elev
external	ext
feet	ft
figure	fig or Fig
finished floor level	FFL
force	F or P
height	H or h
inch	in
internal	int
length	L or l
live load	LL
maximum	max
metre	m
millimetre	mm
minimum	min

not to scale	nts
number	no
pressure	p
radius	R
reinforced concrete	RC or rc
reinforcement	rft
specifications	spec
square	sq
symmetrical	symm
International System of Units	SI
thickness	t
wind load	WL

Glossary

The following are definitions of most of the engineering terms used throughout this book:

Abutment	A supporting pier that receives the thrust of an arch or the reaction of a beam.
Acute angle	An angle less than 90°
Anchor	A metal pin built into concrete or masonry and used to firmly restrain the motion of plates, etc.
Arch	A form of structure with a curved shape.
Bay	The space between two consecutive trusses, frames, beams or lines of columns.
Beam	A supporting member in a structure used mainly to support slabs and walls. Beams are mostly horizontal.
Bolt	A strong metal rod with a head at one end and a screw thread at the other to fit into a nut. Bolts are used in connections to hold separate members together.
Brick	A shaped block of burnt clay used for building purposes.
Bridge	A structure spanning between different locations and over a gap.
Buckling	A form of collapse in which the member deflects sideways suddenly.
Buttress	A concrete or masonry thick member built at a right angle to a wall to stabilize it against earth or water pressure.
Calendering	A means to shape metals by pressing them between revolving cylinders.
Cantilever	A beam or a truss in which one or both ends project beyond the supports.
Cavity wall	A wall built of two leaves of masonry with a continuous space in between.
Channel	A standard rolled steel section with three sides at right angles.
Chord	The top or bottom members of a truss.
Column	A vertical member primarily carrying compression loads.
Column head	An enlargement of the column cross-section used at the connection with the slab in flat slab structures.
Concrete	A stone-like material made by mixing coarse aggregates, sand and cement with water and allowing the product to

	harden. Concrete is typically a stiff material, usually reinforced by steel bars for structural purposes.
Course	A horizontal layer of masonry.
Dead load	A permanent load that consists of the self-weight of the structure and the covering material.
Decagon	A plane geometric shape with ten sides and ten angles.
Diaphragm	A stiff structural member designed to resist forces in its own plane.
Dodecahedron	A regular geometric object with twelve identical faces.
Dome	A shell curved in two directions.
Ductility	The ability to deform under load without fracture.
Elasticity	The property to recover shape after deformation.
Elevation	A drawing that represents an object while being projected on a vertical plane.
Ellipse	A conic section formed by the intersection of a right cone and a plane with the angle between the plane and the cone axis larger than half the core head angle.
Fasteners	Means to fix firmly and securely one member to another.
Fixed support	A support that restrains rotation and displacement in all directions.
Flange	The top or bottom projecting part of an I-beam, channel or Z-beam.
Flat slab	A reinforced concrete slab supported directly on columns without beams.
Footing	A shallow foundation element, normally made of plain or reinforced concrete.
Foundation	A general term describing the base on which a structure rests.
Frame	A skeleton of structural members where the members are connected in hinged or fixed joints.
Girder	A structural member designed to carry mainly bending stresses. Usually a girder carries other beams and/or slabs.
Going	The horizontal interval between consecutive rises in a staircase.
Grout	A mixture of cement, sand and water. Grout is usually pumped or poured to fill voids and hollows in buildings where needed.
Gusset plate	A plate, commonly made of steel, used to connect several members of a truss meeting at a joint.
Handrail	A rail serving as a guard at the side of a bridge or a staircase flight.
Header position	A position of a brick in a brick wall, in which the brick is placed large face down and end-side exposed.
Heptagon	A plane geometric shape with seven sides and seven angles.
Hexagon	A plane geometric shape with six sides and six angles.
Hinged support	A support that allows rotation but restrains displacement in all directions.
Hyperbola	A conic section formed by the intersection of a right cone and a plane with the angle between the plane and the cone axis smaller than half the cone head angle.
Icosahedron	A regular geometric object with twenty identical faces.

Lateral force	A force that acts perpendicular to the axis of line elements, such as beams, or the plane of two-dimensional elements, such as slabs and trusses.
Leaf	A vertical element of masonry with a thickness of one brick.
Links	Thin steel bars used in beams, columns and frames to enclose longitudinal steel bars. Links are also called stirrups.
Lintel	A structural member designed to carry the wall over an opening made to accommodate a window, a door, etc.
Live load	A moving load or variable force on a structure, e.g. the weight of people and furniture in a house.
Marginal beams	Beams used along the edges of a floor to support slabs and walls. They are also called spandrel beams.
Masonry	Stonework, blockwork or brickwork.
Mild steel	Hot-rolled steel with 0.12 to 0.25 per cent carbon, used in manufacturing reinforcement bars and steel structural members.
Mortar	A thick mixture of cement, sand and water. Mortar is used as a binder in masonry construction.
Nail	A slender metal piece with one end enlarged and the other pointed for driving into timber to hold separate pieces together.
Nonagon	A plane geometric shape with nine sides and nine angles.
Obtuse angle	An angle larger than 90°.
Octagon	A plane geometric shape with eight sides and eight angles.
Octahedron	A regular geometric object with eight identical faces.
One-way slab	A slab designed to span between two opposite beams or walls.
Packing plate	A plate used in member joints to fill gaps.
Panel	The portion of a truss between adjacent chord joints. Also a panel is an area of brickwork with defined boundaries.
Parabola	A conic section formed by the intersection of a right cone and a plane with the angle between the plane and the cone axis equal to half the cone head angle.
Pedestal	A short column.
Pentagon	A plane geometric shape with five sides and five angles.
Pier	A compression member formed by a thickened section of a masonry wall.
Pilaster	A part of the wall that projects on one or both sides of its sides. Pilasters are used to add stability to the wall.
Plan	A drawing which represents an object while being projected on a horizontal plane.
Plasticity	The inability to recover shape after deformation.
Plate	A thin, flat piece of metal or other material, usually of uniform thickness.
Polygon	A plane geometric shape with any number of sides.
Purlins	Horizontal members supported on a truss top chord or on a frame. Purlins carry the roof of the building.
Rhomboid	A four-sided geometric plane shape with every two opposite sides equal.
Rhombus	A four-sided geometric plane shape with equal sides.

Ribbed slab	A reinforced concrete slab that has rectangular hollows formed in its bottom side. Usually the hollows are arranged in rows with ribs in between.
Right angle	The angle between two perpendicular lines, 90°.
Rise	The height of a step in a staircase.
Rivet	A metal pin with one head passed through holes in two or more pieces to hold them together. Another head is formed by hammering on the rivet after insertion.
Roller support	A support that allows rotation and displacement in one direction but restrained displacement in all other directions.
Rowlock position	A position of a brick in a brick wall, in which the brick is placed face-side down and end-side exposed.
Screw	A slender metal rod-like with one end enlarged and the stem prepared with a tapering spiral thread. It is driven into timber with the aid of a screwdriver.
Shear wall	A vertical wall designed to resist horizontal, in-plane forces such as wind loads.
Sheet	A broad and thin piece of metal or other material. A sheet is typically thinner than a plate.
Shell	A thin curved surface.
Slab	Normally a reinforced concrete floor supported on beams, columns and/or walls, usually of uniform thickness.
Slab drop	A slab additional thickness, used around the connections with the columns in flat slab structures.
Soldier position	A position of a brick in a brick wall, in which the brick is placed end-side down.
Solid slab	A reinforced concrete slab with a solid section.
Spacing	The distance between each two successive objects in a row.
Span	The distance between the supports of a structural element.
Spandrel beams	See marginal beams.
Splice plate	A plate used to joint pieces of metal, timber or other material.
Stair	One step in a series forming a means of passage from one floor to another in a building.
Staircase	A flight, or a series of flights, of stairs with its framework.
Steel	A modified form of iron, produced artificially to be suitable for structural use.
Steel reinforcement	Steel bars or meshes added to concrete structural members to increase their strength.
Stiffener	A structural strengthening, such as an angle, fixed to a plate to prevent its buckling.
Stirrups	See links.
Stone	The hard natural substance of which rock consists.
Storey	A complete horizontal section of a building having one continuous floor.
Strength	The ability to resist forces.
Stretcher position	A position of a brick in a brick wall, in which the brick is placed large face down and long thin side exposed.
Timber	The type of wood that is suitable for building purposes.
Toe board	A timber board fitted along the sides of bridge decks to prevent the slipping of objects out of the deck.

Trapezium	A four-sided goemetric plane shape with all sides and angles different.
Trapezoid	A four-sided geometric plane shape with two opposite sides parallel.
Truss	A steel or timber framework designed to cover moderate to large spans, and in which the members carry only tension or compression stresses.
Two-way slab	A slab designed to be supported along all four sides.
Wall	A vertical planar structural element, designed to carry vertical and sometimes also lateral loads.
Washer	A circular metal ring used under bolt heads and nuts to distribute the bearing load from the bolt to the connected members over a large area.
Web	The portion of an I-beam, channel or Z-beam that extends between the flanges.
Web members	The truss members that extend between the top and bottom chords.
Welding	Uniting pieces of metal after being melted by heat.
Wind load	The force acting on a structure due to the pressure of wind.
Wood	The hard natural material composing most of tree trunks and branches.

PART 1

Basics of Drawing

Part 1 of this book is concerned with the basics of drawing. It introduces engineers and draughtsmen to the main two types of projection drawing, namely orthogonal and pictorial. This part of the book starts with an introduction to drawing and then covers the use of drawing instruments with emphasis on engineering applications, the use of lines and the art of lettering. The book also includes a chapter on simple drawing operations—the basic skills every draughtsman should perfect. Among the sections covered in this chapter are sections on conic sections and mathematical curves. These sections are intended to be for reference purposes.

The book then discusses orthogonal projection and two types of pictorial projection—isometric and oblique—in three consecutive chapters. Every projection technique is explained, starting from the projection of points, then lines and surfaces and finally objects. Hints to make projection convenient and easy are presented, and examples are used all the time to assist learning. The disadvantages of every projection technique are not masked but are put forward in detail so that the reader understands both the merits and shortfalls of every technique. The reader is strongly advised to study these sections on the evaluation of projection techniques.

Part 1 of the book ends with a chapter on the intersection of geometric objects. This chapter is particularly important in the field of civil engineering applications where objects of different shapes intersect. Although this part of the book is aimed at general drawing techniques, special emphasis is given to civil engineering applications whenever possible.

A large number of exercises is included in every chapter. They are arranged in an ascending order of difficulty. Although some of the exercises appear to be simple, they are in fact difficult and require a great deal of practice and imagination. Nevertheless, they have been designed to help assess the understanding gained through reading this book and can be quite exciting.

CHAPTER 1 Introduction

1.1 General introduction

Engineering drawing is an international language which expresses itself through lines instead of words. It enables ideas to be expressed and communicated in an easy and clear way. The language of engineering drawing deals with information, normally related to shapes, that can not be conveyed in words and letters. To achieve this purpose, the language of drawing has an established standard system of rules and conventions, which are truly international. Like any other language where writing and reading are the essential means of communication, presentation and visualization are those that correspond to drawing. Representation of a shape is done by means of lines so that they can quickly construct a visual image of the shape in the receiver's mind. To illustrate the superiority of drawing over verbal description in representing shapes, see Figs 1.1, 1.2 and 1.3.

The representation of three-dimensional objects or shapes on two-dimensional surfaces by drawing has evolved from a long gradual change through centuries. One of the earliest engineering drawings that exist today is an Egyptian drawing made on papyrus for two views of a shrine without dimensions.

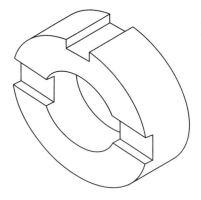

Figure 1.1 A mechanical part

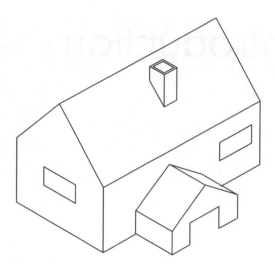

Figure 1.2   A bungalow

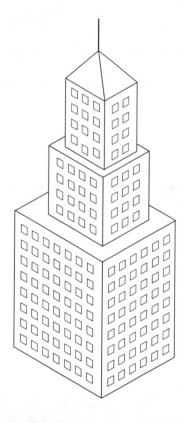

Figure 1.3   A high rise building

1.2 Purposes of engineering drawing

All engineering and architectural objects require various kinds of drawings:

■ Preliminary drawings to assist preliminary design and to facilitate supplying the computer with the input data of the structure;
■ Layout drawings to position structural parts in their relative locations;
■ Detailed drawings of structural members and their interconnections.

Some of the common uses of drawings are stated in the following.

Design of buildings

Typical drawings of building construction represent foundation details, floor plans for every storey and sectional views of various structural members and connections. Other drawings are needed for the water and electricity supplies, plumbing, heating and ventilation systems, window and door details, etc.

Machine design

After a machine or a mechanical part is designed, detailed drawings are prepared to describe it for production. This is normally associated with verbal instructions and specifications. Drawings to describe the size and location of each part in a machine are also needed to facilitate the assembly of the machine.

Roadway and railway projects

Firstly, maps showing the exact location of the road and the boundary lines of neighbouring properties are essential. Also, longitudinal sections through the road are required to specify road levels. Frequently, cross-sections at intervals are drawn to depict the fill and/or cut of the surrounding earth and the details of the surfacing layers.

There are many other uses of engineering drawings such as in defence installation projects, aerospace applications, etc.

1.3 Projection

Two types of projection drawings are normally used in the fields described in Sec. 1.2, namely orthogonal and pictorial. In either case, drawings must be precise and complete.

Orthogonal projection

Orthogonal projection is commonly preferred for the presentation of manufacturing and construction drawings. This method represents a shape by means of a set of two-dimensional views, usually a view from the top, a plan, and two views from two sides, front and side elevation. In this method, complex shapes are fully described. However, orthogonal projection does not create an immediate picture of the shape in the reader's mind. It takes training and practice in both the preparation and the interpretation of orthogonal views to enable the formation of a quick mental picture of the shape. For a complete explanation, refer to Chapter 5.

Pictorial projection

Pictorial projection views actually create the illusion of being three dimensional. Therefore, they possess the ability to create an immediate picture of the shape in the reader's mind—even the inexperienced one. However, they carry the drawback of being incomplete with some details lost. For example, the depth of the hole in Fig. 1.4 is not given. Pictorial views are usually used in architectural drawings to demonstrate the three-dimensional quality of a building, such as those shown in Figs 1.2 and 1.3. An introduction to pictorial projection is presented in Chapter 6. The two types of pictorial projection most commonly in use in civil engineering applications, namely isometric and oblique, are discussed in Chapters 6 and 7.

1.4 What is good drawing?

Drawing is the main language of communication between the designer and the constructor. The following recommendations are introduced as keys to good drawing:

1   Unnecessary details should be avoided. A drawing should only convey the information for which it is required.

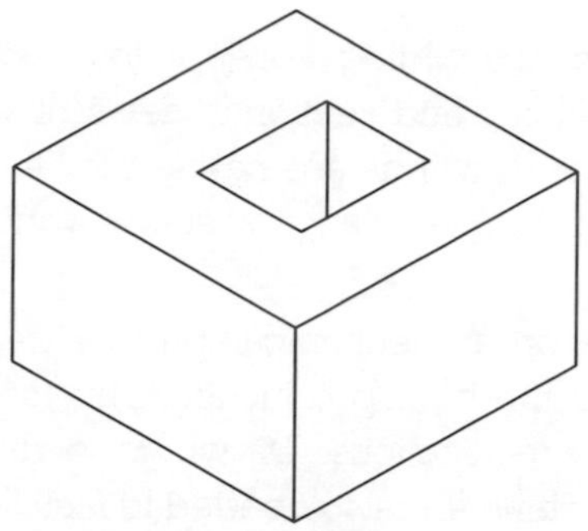

Figure 1.4 A geometric object

2   Drawings must be clear with no possibility for wrong interpretation.

3   Drawing scales should be chosen so that the drawing is not too big for easy handling and the details are not too small for proper recognition.

4   Use as few scales as possible on each drawing. A scale should be chosen so that easy drawing and interpretation are allowed. Scales of 1:2, 1:5 and 1:10 are easier to work with than 1:4 and 1:8. Scales such as 1:3, 1:7 and 1:9 should be avoided.

5   Instructions and notes must be decisive. Comments such as 'The concrete may be compacted' or 'The bars could be joined' must not be used; instead use 'Compact concrete' and 'Join bars'.

6   Every drawing should have a self-explanatory title and description, and must be complete in itself. Cross reference between related drawings should be made to facilitate identification of drawings for a particular job.

7   Drawing standards and conventions must be followed in all drawings and must be consistent within a whole project.

CHAPTER 2 Drawing Instruments

2.1 Introduction

Engineering and architectural drawing is an art as much as a science. Drawings must not only be accurate, precise and complete, but also neat and well organized. For this purpose, engineers and draughtsmen should learn to use the most appropriate instruments in the most appropriate manner. This also helps reduce drawing time.

The drawing instruments used regularly include: a drawing board, drawing sheets, a T-square, pencils, triangles, a compass, a divider, a graduated ruler, irregular curves, a protractor and an eraser. In the following Secs 2.2 to 2.10, the use of each of these instruments is discussed.

2.2 Drawing board

This is usually a timber board with a smooth and absolutely flat surface. It has a straight working edge on the left-hand side, sometimes fitted with a metal or a plastic shield. The working edge is for the T-square to slide on with a true right angle between the T-square and this edge.

2.3 Drawing sheets

There are various standard sizes of drawing sheets available: A0 (841 mm × 1189 mm), A1 (594 mm × 841 mm), A2 (420 mm × 594 mm), A3 (297 mm × 420 mm) and A4 (210 mm × 297 mm). The choice of a size depends on the size of drawing in hand. Normally, A0 and A1 sizes are used in building and construction projects.

A drawing sheet is held on a drawing board by means of masking tape or sellotape. Short pieces of tape are used at the corners of the sheet with some tension applied to the tape to keep the sheet flat, as shown in Fig. 2.1. Staples and pins should be avoided as they damage the drawing sheet and drawing board.

Each drawing sheet should have a title panel and an information panel, normally positioned in the right-hand side of the sheet (see Fig. 2.2). In case of limited space, the title and information panels may be positioned in the bottom strip of the sheet. The information panel typically includes the following information in a descending order from top to bottom:
- Notes, such as specifications and recommendations, and
- Revisions and amendments, along with the name of person approving each and the corresponding date.

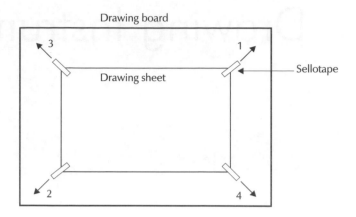

Figure 2.1   Drawing sheet held on a drawing board by sellotape; 1 to 4 show the steps of fixing the drawing sheet

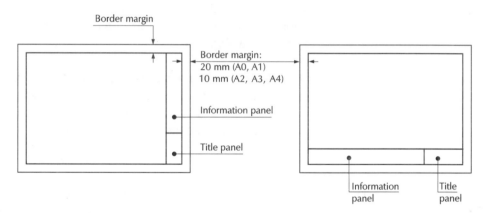

Figure 2.2   Title panel and information panel

The title panel contains the following items:
- Project title,
- Title and number of drawing,
- Scales used,
- Date of drawing,
- Name of designer,
- Name of client and
- Identification of persons carrying out the draughting and checking.

2.4 T-square

The common T-square consists of two parts fixed together: the head which slides along the working edge of the drawing board and the blade which is used to draw horizontal lines and to support triangles and set-squares (see Fig. 2.3). Various lengths of blade are available, the choice of which depends on the size of the drawing sheet to be used. The blade must not be used as a guide for a cutter as this will damage its straight edge.

2.5 Pencils

Special hexagonally shaped pencils are used in drawing. Various degrees of softness, and consequently blackness, are available, ranging from the hardest, 9H, to the softest, 7B (see Fig. 2.4).

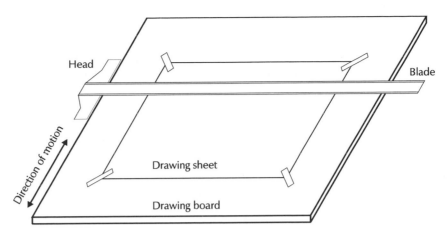

Figure 2.3 T-square

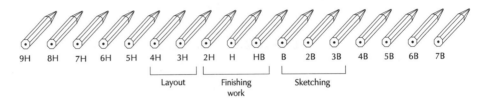

Figure 2.4 Drawing pencils

Clutch pencils are now also available. They can produce any degree of softness according to the lead used. Leads are available in different diameters: 0.3, 0.5, 0.7 mm, etc. They produce lines of constant thickness and do not require sharpening.

Generally, 3H and 4H pencils are appropriate for layout work, while HB, H and 2H pencils are most useful in making a finished pencil drawing. For sketching, B and softer pencils are adequate.

In drawing straight lines, the pencil ought to be held at 50 to 60° to the drawing sheet and in a plane perpendicular to it. The proper directions of drawing straight lines are illustrated in Figs 2.5 and 2.7.

Figure 2.5 Drawing straight lines

2.6 Triangles The most common types of triangles are 45-45-90° and 30-60-90° triangles, with which angles of 15° plus any multiplies of 15° can be directly drawn. For other angles, adjustable set squares are used (see Fig 2.6). Figure 2.7 shows how the standard triangles are used to produce lines at different angles and the drawing directions associated with them. Triangles are also used to draw parallel and perpendicular lines, as shown in Fig. 2.8.

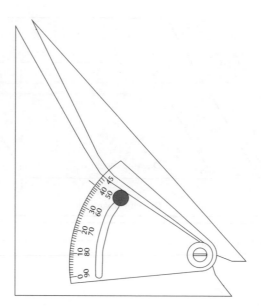

Figure 2.6   Adjustable set square

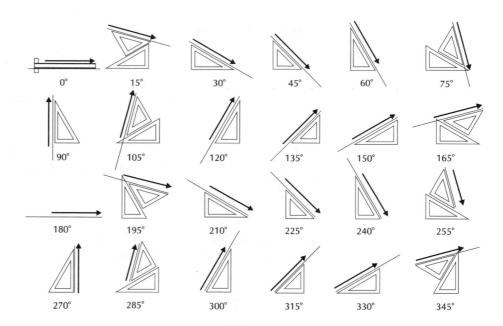

Figure 2.7   Using triangles to draw inclined lines

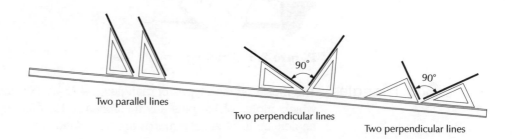

Figure 2.8   Drawing parallel and perpendicular lines using triangles

2.7 Compass A compass is used to draw circles and arcs. It has two legs, one ending with a needle point to sink into the paper and the other ending with a lead to draw arcs (see Fig. 2.9). This latter leg may be modified in some cases as follows:

1 It may accommodate an extension beam to draw large arcs.

2 It may be replaced with an inking end for inking purposes.

3 The leg may be replaced with another with a needle point end. In this case, the compass can be used as a divider (see Sec. 2.8).

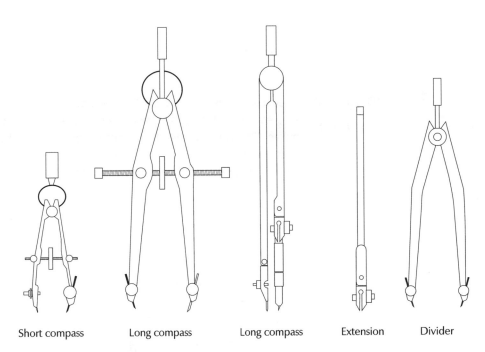

Short compass   Long compass   Long compass   Extension   Divider

Figure 2.9 Compasses

2.8 Divider Dividers are used regularly to transfer distances. To set a divider, it should be opened larger than the distance to be transferred and then closed down to the proper measurement by pressing on one of its legs with the thumb. Once the divider is set, it can be used on the drawing sheet to mark two points separated by the measurement previously taken.

2.9 Graduated rulers and scales Rulers are available in different lengths, e.g. 100, 200, 300 mm, etc. They are normally graduated in millimetres and centimetres, although inch graduation is sometimes used.

Scales, on the other hand, are rulers with a scale included in the graduation. They facilitate drawing at a scale while removing the need to adjust dimensions through scale during the process of drawing (see Fig. 2.10).

To mark off a dimension with a ruler or a scale, a very sharp pencil or a needle point must be used and held vertically to avoid discrepancy.

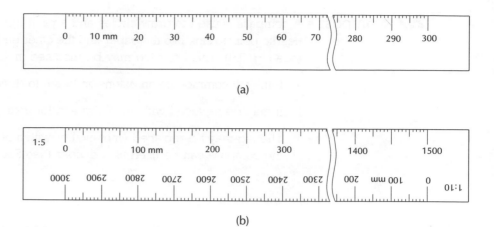

(a)

(b)

Figure 2.10   (a) A graduated ruler and (b) a scale

2.10 Irregular curves

Irregular curves (also called French curves) are used to draw non-circular arcs as they can not be drawn with a compass. Figure 2.11 shows a typical irregular curve. To draw a curve, the irregular curves may be used several times to draw parts of the curve step by step, as shown in Fig. 2.12. It is of utmost importance to avoid any humps or kinks between the different steps, and this certainly needs practice.

Figure 2.11   Irregular curve.

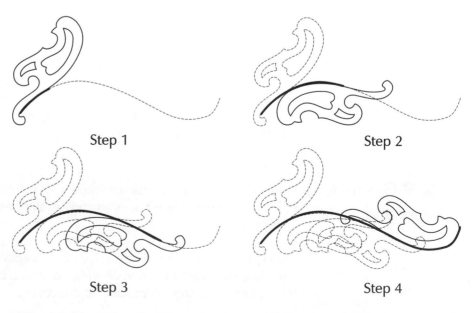

Step 1   Step 2

Step 3   Step 4

Figure 2.12   The use of irregular curves to draw curved lines

Flexi-curves may also be used to draw non-circular arcs. A flexi-curve is composed of a plastic tube with a rectangular cross-section, in which a flexible metallic rod is inserted. When deforming a flexi-curve to make it follow a certain curve, the metallic rod inside allows it to retain the shape. However, in practice, shaping a flexi-curve is often difficult as deforming one part may affect the shape of other parts. Also, shaping it to follow a curve with sharp corners may prove to be impossible. In general, the use of flexi-curves to draw non-circular curves in accurate work is not recommended.

2.11 Important notes

It can not be overemphasized that drawing is an art as well as a science. Drawings must be clean and neat as well as precise and complete. It is a fact of life that once techniques (whether they belong to drawing or not) are *learned* incorrectly, it is very difficult to *unlearn* them and then *relearn* correctly. For this reason it is recommended that a beginner should observe the following suggestions.

1 Keep hands off the drawing sheet. A small piece of paper or cloth could be used to rest hands on during drawing.

2 Triangles are better picked up than slid over the sheet.

3 The head of the T-square should be slightly pushed down to raise the blade before sliding the T-square over the drawing sheet.

4 Hard pencils (3H or harder) are better used for layout work.

5 Pencils should never be sharpened on the drawing sheet.

6 Drawing instruments must be cleaned regularly before and during drawing.

2.12 Exercises

1 Draw to a proper scale the house plans shown in Figs 2.13 and 2.14. All dimensions given are in metres.

2 Draw to a proper scale the survey traverse network lines shown in Figs 2.15 and 2.16.

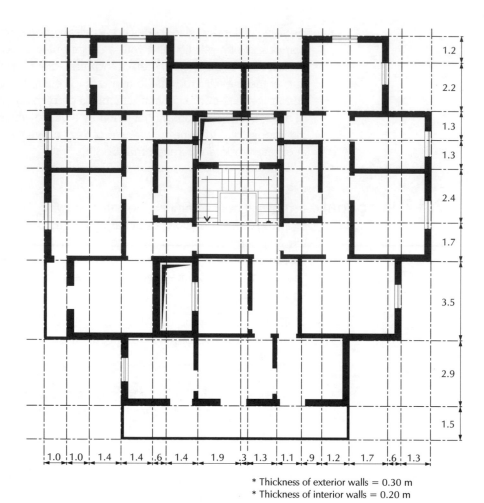

* Thickness of exterior walls = 0.30 m
* Thickness of interior walls = 0.20 m

Figure 2.13

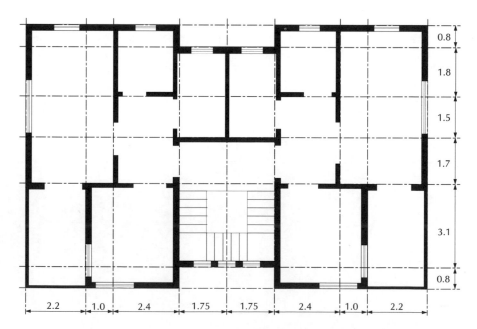

* Thickness of exterior walls = 0.30 m
* Thickness of interior walls = 0.20 m

Figure 2.14

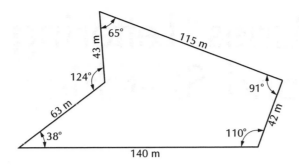

Figure 2.15

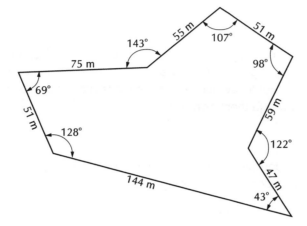

Figure 2.16

CHAPTER 3

Lines, Lettering and Sketching

To produce neat drawings, special requirements for drawing lines and for lettering should be observed. Inexperienced engineers and draughtsmen who do not follow these requirements may cause confusion or create wrong ideas through their drawings.

3.1 Lines

According to British Standard specification BS1192, the following line styles may be used in engineering and architectural drawings (see Fig. 3.1).

(a) Continuous thick lines for visible outlines and edges.

(b) Continuous thin straight or curved lines for dimension and leader lines and hatching.

(c) Thick or thin dashed lines for hidden outlines and edges.

(d) Chain thin lines for centre lines, symmetry lines and reference lines.

(e) Cutting plane lines for cutting planes. They are chain thin lines, but thick at ends and changes of direction.

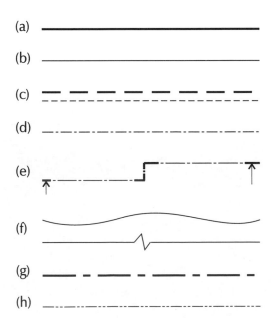

Figure 3.1 Line styles used in engineering and architectural drawings

(f) Break lines. These can be continuous, thin and drawn freehand, or continuous, thin and with a zigzag. These lines are used for limits of partial or interrupted views and sections provided that the limit is not an axis; otherwise a chain thin line is used.

(g) Chain thick lines used as indication of lines or surfaces to which a special requirement applies.

(h) Chain thin double-dashed lines used as outlines of adjacent parts.

The descriptions, thick and thin used above are used in relative terms. The actual thickness of lines in a drawing depends on its purpose, size and scale, and whether the lines are drawn in ink or in pencil. For a 1:1 scale, the minimum line thickness recommended by BS1192 is 0.25 mm. The line thickness should be consistent in all drawings of any one project.

Figure 3.2 shows the correct way to draw dashed lines and centre lines, and some common errors. The use of line styles in drawing views of a geometric object is illustrated in Fig. 3.3.

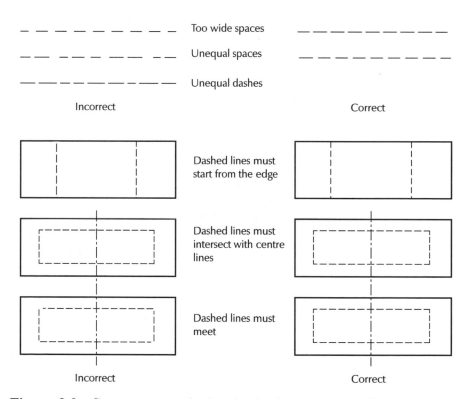

Figure 3.2 Common errors in drawing broken and centre lines

3.2 Dimension lines

Although geometric drawings should be drawn to a scale, direct measurement from the drawings is not recommended, and instead all required dimensions must be explicitly given. The dimensions given for a geometric shape should identify all its details in all three directions. They should also give the location of each part in relation to other parts. A dimension consists of the following elements:

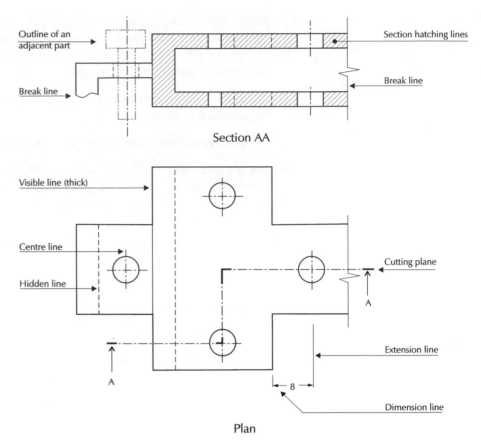

Figure 3.3   Two views of a geometric object

1 *Projection or extension lines (sometimes called witness lines)*   These are light lines drawn at the ends of, and usually perpendicular to, the line for which a dimension is needed. They should not touch the shape and a 1 to 2 mm space is left in between. Sometimes, extension lines are not drawn where other lines such as symmetry lines exist. Also, if the dimension line is drawn inside the shape, extension lines are not required (see Fig. 3.4). Extension lines have the same weight as dimension lines.

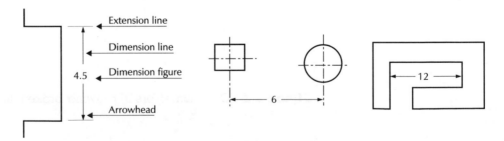

Figure 3.4   Extension lines

2 *Dimension line*   This is a light line, normally drawn parallel to the line for which a dimension is required, unless an inclined dimension is needed (see Fig. 3.5). Dimension lines are normally placed with a uniform spacing of 6 to 10 mm.

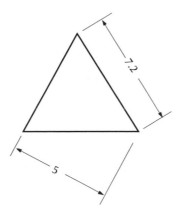

Figure 3.5 Dimension lines

3 *Arrowheads* Arrowheads are drawn freehand as neatly as possible. They consist of two edges of a triangle with a 15° head angle. In small drawings, arrowheads may be solid. As alternatives to arrowheads, a dimension line may end with short oblique strokes, dots or small circles (see Fig. 3.6). However, it is recommended that only one type of dimension line termination be used on any one drawing.

Figure 3.6 Styles of dimension line termination: (a) arrowheads, (b) oblique strokes, (c) solid dots, (d) small circles

4 *Figures* Figures are written clearly with heavy lines, either above the dimension line (the structural practice) or in a space in the dimension line (general practice). Figures may be written horizontally or written such that they can be read from the bottom right corner of the drawing sheet. They are commonly given in SI units. Although the International System of Units is now commonly in use, it should be stated, as one of the notes, that 'All dimensions are in millimetres' (or in metres).

Dimensions are required not only for straight distances but also for angles, arcs, curves, etc.

1 *Straight distances* Dimensions for straight distances can take one of the forms shown in Fig. 3.7 according to the space available. In some cases it might be appropriate to dimension a series of sequential distances that are equal to each other using one dimension line, as shown in Fig. 3.8. In such cases, the number of distances, the length of every distance and the overall length are given.

Figure 3.7 Dimensioning of straight distances

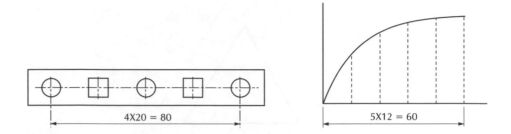

Figure 3.8 Dimensioning a series of sequential distances

2 *Angles* A circular arc, whose centre is at the angle head, is drawn, and the angle magnitude in degrees is written horizontally where the space allows (see Fig. 3.9).

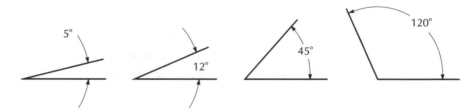

Figure 3.9 Dimensioning of angles

3 *Curves* Dimensions for curves may be given by means of offsets, as shown in Fig. 3.10a, or by coordinates relative to two perpendicular, or non-perpendicular, axes (see Fig. 3.10b).

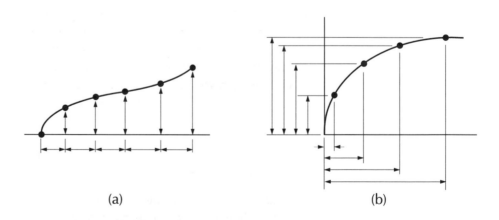

(a)   (b)

Figure 3.10 Dimensioning of general curves

4 *Small circles* There are various ways to give the dimensions of small circles. Figure 3.11 shows some of these for circles on a straight line or on the perimeter of a large circle. Rectangular or polar coordinates may be used.

When pictorial drawing is considered, the above rules apply in addition to some special requirements discussed in Secs 6.7 and 7.8.

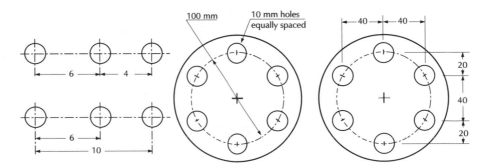

Figure 3.11 Dimensioning of small circles

Important notes   To avoid confusion in reading the dimensions of geometric shapes, the following notes should be observed:

1 Dimensions should be put outside the drawing unless it is necessary to add some inside (see Fig. 3.12).

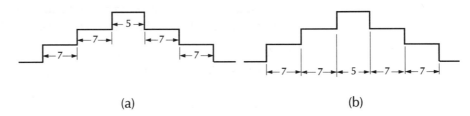

(a)   (b)

Figure 3.12   (a) Not recommended and (b) appropriate methods of dimensioning

2 Dimensions that are not needed or can be obtained from other given dimensions should be discarded.

3 A dimension should be used in the view or direction that gives the true magnitude. For example, a circle may appear as an ellipse in some views. In this case, the radius should be given in the direction that gives its true magnitude (see Fig. 3.13).

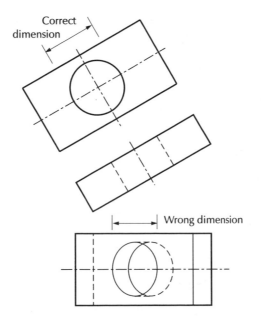

Figure 3.13   Correct dimensioning of a circle in an inclined plane

4   Scattering dimension lines should be avoided. It is better, and easier to read, to put them all in one or two lines.

5   Dimensions are arranged such that inner dimensions are closest, and outer dimensions are farthest, from the shape (see Fig. 3.14).

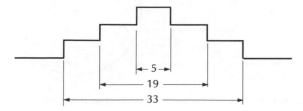

Figure 3.14 Arranging dimensions

6   A dimension line should not intersect with an extension line other than its own lines.

7   Dimension lines must not be packed in a small space. In the worst case, a complex part of the drawing may be reproduced at a larger scale to enable a better distribution of dimension lines (see Fig 3.15a).

8   Where it is necessary to put a dimension line inside a hatched area, a space must be left blank for the dimension figure, as shown in Fig. 3.15b.

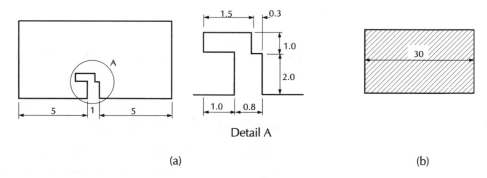

Detail A

(a)   (b)

Figure 3.15 (a) Dimensioning of small details and (b) dimension figure in a hatched area

9   In pictorial views, extension lines should be drawn in the same plane as the face for which the dimensions are given (see Fig 3.16).

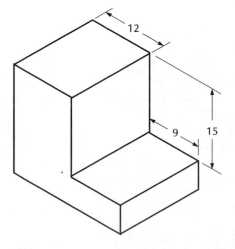

Figure 3.16 Pictorial view of a geometric object

As an example, Fig. 3.17 presents an incorrect and a correct dimensioning system of a geometric object.

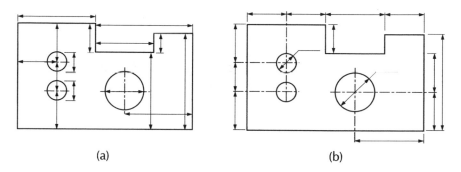

(a)   (b)

Figure 3.17   (a) Incorrect and (b) correct dimensioning systems

Leader lines   Leader lines link drawing details with corresponding notes (see Fig. 3.18). They may end:

■ With a dot, if they end inside an object,
■ With an arrowhead, if they end on the outline of an object, or
■ Without a dot or arrowhead, if they end on a dimension line.

The angle between leader lines and dimension lines or object outlines should be steep, i.e. more than 60°.

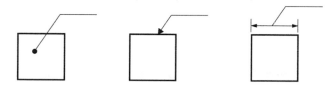

Figure 3.18   Leader lines

3.3 Letters and numerals

Written notes are used to convey information that can not be easily given graphically. They should be grouped and placed as near as possible to the part they refer to and not spread over the drawing. It is also essential for the production of a neat drawing to do the lettering in an organized and concise form. It is always considered that lettering is a skill that every draughtsman must perfect. Lettering must be easy to read and understand as bad freehand lettering may corrupt a nicely drawn drawing. Practice is necessary to write neatly and consistently.

Stencils are available as aids to lettering. However, these should not be resorted to unless necessary. A good freehand lettering style is more pleasing and quicker than use of a stencil.

Underlining of notes on a drawing is not generally recommended.

Lettering styles   The commonest lettering style is Gothic. Roman and Text (such as Old English) are sometimes used. The Gothic style is the simplest style with all parts of letters having the same thickness. The Gothic lettering style is shown in Fig. 3.19. Writing in Gothic can be in small or capital letters, and in normal (vertical) or italic letters (inclined at 15° to 25° to the vertical) (see Fig. 3.20).

a b c d e f g h i j k l m n o p q r s t u v w x y z
A B C D E F G H I J K L M N O P Q R S T U V W X Y Z
1 2 3 4 5 6 7 8 9 0

Figure 3.19   Gothic letters

a b c d e f g h i j k l m n o p q r s t u v w x y z
A B C D E F G H I J K L M N O P Q R S T U V W X Y Z
1 2 3 4 5 6 7 8 9 0

Figure 3.20   Italic style of letters

Size of letters   The given size of lettering is normally the height of capital letters and numerals. The height of small letters is two-thirds of that height. Some small letters, such as b, d and h, have stems that are as high as the capital letters. Some small letters, such as g and y, have descenders that extend one-third of the height of the capital letters below the bottom line. See Fig. 3.21 for examples of these letters. Table 3.1 gives the minimum heights of letters and numerals recommended by BS1192.

h = height of capital letters

Figure 3.21   Height of letters and numerals

Table 3.1   Minimum height of letters and numerals

Application	Size of drawing sheet	Minimum height (mm)
Titles, drawing numbers, etc.	A0, A1, A2 and A3	7
	A4	5
Dimensions and notes	A0	3.5
	A1, A2, A3 and A4	2.5

The mechanics of lettering   It is said that 'practice makes perfection'. This is absolutely true with regard to the production of high-quality lettering. Spacing between, and shape of, letters are important factors in producing clear and neat lettering, as explained in Fig. 3.22.

W i d e s p a c i n g and narrowspacing are hard to read.
Even spacing is important. Un even sp a cingisnot accept able.
All letters should have the same slope. Otherwise the appearance is spoilt.

Figure 3.22   Spacing of letters

It is advisable that lettering is done within predetermined guidelines, as shown in Fig 3.21. Four guidelines are required for every line of lettering: two to guide capital letters, one to guide the top of small letters such as a, c and e and one to guide the descenders of small letters such as g, j and y. Guidelines are drawn lightly so that they do not show on drawing prints and photocopies.

Another common advice is to draw letters and numerals from the top downwards. As an example, the directions of drawing capital letters are shown in Fig. 3.23.

Figure 3.23 Directions of drawing capital letters

3.4 Sketching

Sketching is the ability to draw neatly and accurately without instruments. It is an important skill a professional engineer should develop. On many occasions, an engineer finds it necessary to draw free-hand sketches to express ideas and communicate with a draughtsman in a design office or with a fellow engineer on site. Sketches may also be used for other purposes, e.g. to construct a pictorial view of an object, for which only orthogonal views are given. This helps visualize the drawn object.

It is not necessary to be an artist to draw good free-hand sketches. It only requires practice to draw without drawing instruments. However, it is a well-known fact that acquiring proficiency in free-hand sketching is more difficult and requires more effort than proficiency in drawing with instruments. For example, drawing a straight line, two parallel lines or a circle can be easily done with instruments, but it is a different matter to draw them neatly freehand (see Fig. 3.24).

Free-hand sketching is used to draw orthogonal and pictorial views following the same rules as for drawing with instruments.

Materials required for sketching

Only a pencil, a sheet of paper and an eraser are required in free-hand sketching. Normally, a soft pencil, B or 2B, is used. Blank paper is often used although sometimes paper with printed grids may be used. Grids may be rectangular for sketching orthogonal and oblique views or have lines at 0, 30, 90 and 150° for sketching isometric views.

Mechanics of sketching

Although sketching is a skill which requires the judgement of draughtsmen in controlling the dimensions, straightness of lines, curvature of circles, etc., a few guidelines may still be set up. These guidelines are based on the experience of professionals and, if followed, they may help beginners acquire a high standard of free-hand sketching skill.

1 Hold the pencil at a high level, about 10 mm higher than the position held for normal writing and drawing.

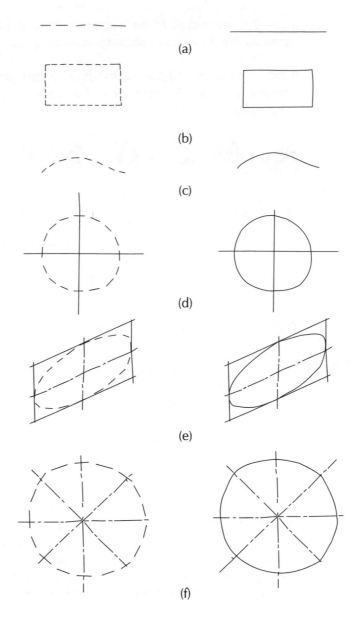

Figure 3.24 Freehand sketching of (a) straight line, (b) rectangle, (c) curve, (d) small circle, (e) ellipse, (f) large circle

2 If intermediate points can be located, mark a sufficient number of these to make drawing the line accurate.

3 Start with dashed light lines (and not full lines).

4 In the case of inclined lines, it is advisable to rotate the sketching sheet so that inclined lines can be drawn in a horizontal position.

5 Judge the proportions of the sketch and move lines if required.

6 Judge the straightness and curvature of line and make any necessary modifications with dashed light lines.

7 Finally, all lines are drawn with heavy lining.

Figure 3.24 illustrates the sketching of a straight line, a box, a curve, a small circle, a large circle and an ellipse. Figures 3.25 and 3.26 show free-hand

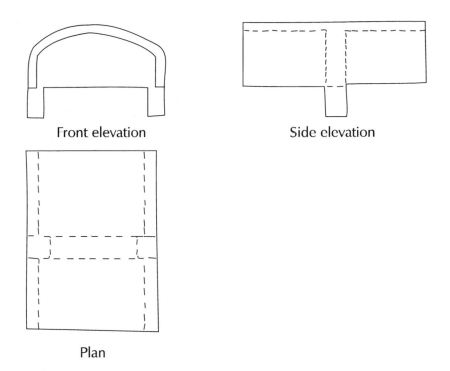

Front elevation   Side elevation

Plan

Figure 3.25   Orthogonal views of a double-cantilever shell

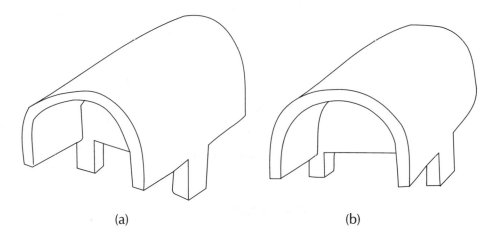

(a)   (b)

Figure 3.26   (a) Isometric and (b) oblique views of a double-cantilever shell

sketches of a double-cantilever shell in orthogonal, isometric and oblique projection views.

3.5 Exercises

1   Find the errors in the lining and dimensioning of the drawings presented in Figs 3.27 to 3.30.

2   Draw the objects shown in Figs 3.31 to 3.36 and illustrate how each object can be fully dimensioned.

3   Draw free-hand sketches of the objects shown in Figs 3.37 to 3.45.

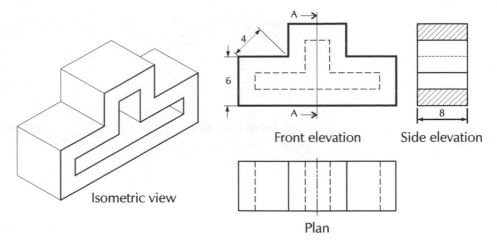

Isometric view

A →

4

6

A →

Front elevation

Side elevation

8

Plan

Figure 3.27

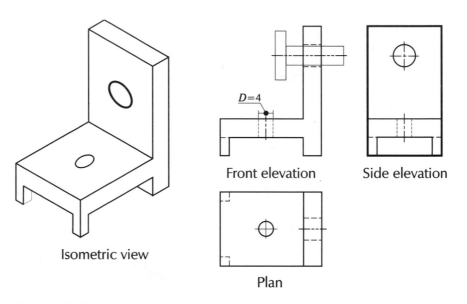

Isometric view

$D=4$

Front elevation

Side elevation

Plan

Figure 3.28

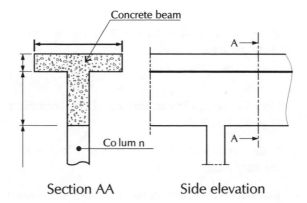

Concrete beam

A →

Column

A →

Section AA

Side elevation

Figure 3.29

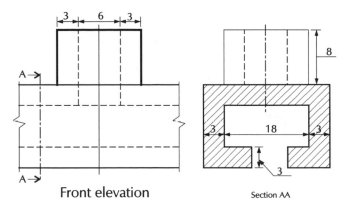

Front elevation   Section AA

Figure 3.30

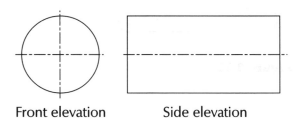

Front elevation   Side elevation

Figure 3.31

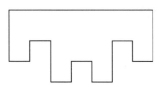

Figure 3.32

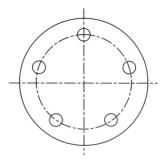

Figure 3.33

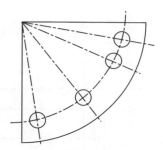

Figure 3.34

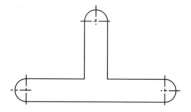

Figure 3.35

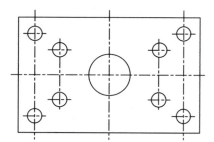

Figure 3.36

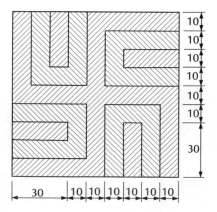

Figure 3.37

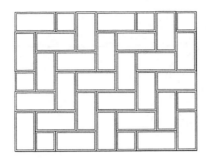

Figure 3.38

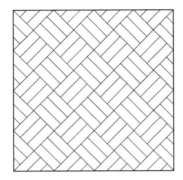

Figure 3.39

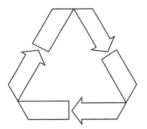

Figure 3.40

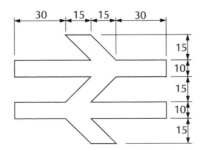

Figure 3.41

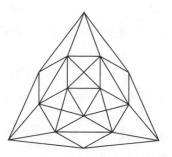

Figure 3.42

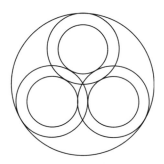

Figure 3.43

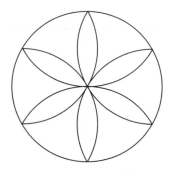

Figure 3.44

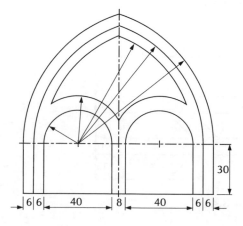

Figure 3.45

CHAPTER 4

Simple Drawing Operations

4.1 Introduction

There are a few simple and basic drawing operations that are used extensively in engineering and architectural drawing. They are so important that not knowing them will surely hinder the drawing process. These operations are discussed in this chapter with self-explanatory examples.

4.2 Dividing a straight line into a number of equal parts

The easiest way to divide a line into a number of equal parts is by means of a ruler. However, if a higher accuracy is required the methods illustrated below may be used.

Dividing a line into two halves

A straight line can easily be divided into two halves using a compass as shown in Fig. 4.1.

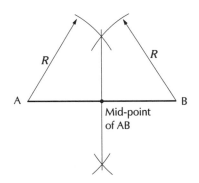

Figure 4.1 Dividing a line into two halves

Dividing a line into an even number of equal parts

A 45–45–90° or a 30-60-90° triangle may be used to divide a straight line into an even number of equal parts, as shown in Fig. 4.2.

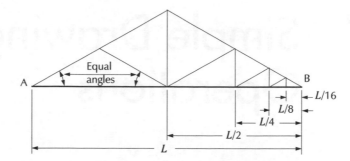

Figure 4.2   Dividing a line into an even number of equal parts

Dividing a line into three
(or multiplies of three)
equal parts

This can be done using a 30–60–90° triangle as shown in Fig. 4.3.

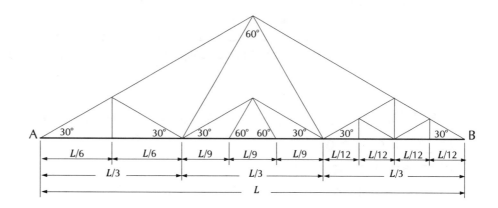

Figure 4.3   Dividing a line into multiples of three equal parts

Dividing a line into any
number of equal parts

To divide line AB (Fig. 4.4) into, say, five equal parts:

1   Draw another line AC at an appropriate angle to AB and with an appropriate length.

2   Mark five equal lengths on AC using a divider to position points D, E, F, G and H.

3   Join H and B and draw parallels to HB at D, E, F and G to divide AB into five equal parts, AI, IJ, JK, KL and LB.

Alternatively, dividing a line into a number of equal parts may be done using a ruler touching or slightly away from the line, as shown in Fig. 4.5a and b. The same technique may be applied to divide a line into a number of unequal parts.

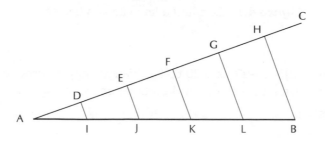

Figure 4.4   Dividing a line into five equal parts

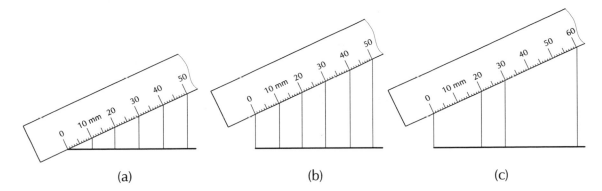

(a)   (b)   (c)

Figure 4.5 Dividing a line into a number of equal or unequal parts

For example, the line shown in Fig. 4.5c is divided into three parts with the ratio 2:1:3.

4.3 Drawing circles

Circles can be drawn if known to be passing through at least three points, or tangent to at least three lines.

Drawing a circle that passes through three points

To draw a circle that passes through three points, the steps given below must be followed. Refer to Fig. 4.6.

1 Join the three points with straight lines to form a triangle.

2 Draw a perpendicular to every line at its mid-point.

3 The point of intersection of the three perpendiculars is the centre of the circle. If the accuracy of drawing is guaranteed, only two perpendiculars need be drawn.

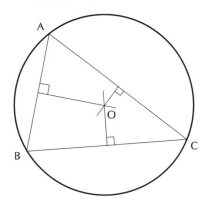

Figure 4.6 Drawing a circle that passes through three points

Drawing a circle that is tangent to three (or more) lines

As a rule, lines that divide angles between the tangents of a circle meet at the centre of the circle (Fig. 4.7).

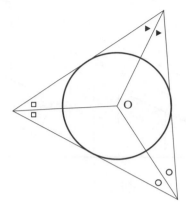

Figure 4.7 Drawing a circle that is tangent to three lines

4.4 Drawing regular polygons

Any geometric figure enclosed entirely by straight lines is called a polygon. Regular polygons have equal sides and equal interior angles. Among the regular polygons are the equilateral triangle (three sides), the square (four sides), the pentagon (five sides), the hexagon (six sides), the heptagon (seven sides), the octagon (eight sides), the nonagon (nine sides) and the decagon (ten sides). In irregular polygons, some or all sides have different lengths. Among the irregular four-sided polygons are the rectangle, the rhombus, the rhomboid, the trapezoid and the trapezium (see Fig. 4.8).

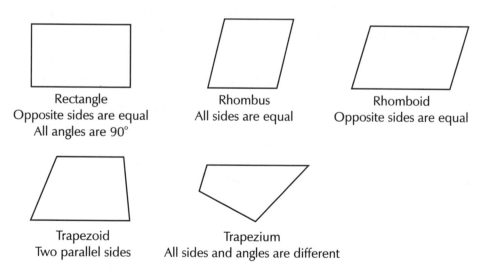

Rectangle
Opposite sides are equal
All angles are 90°

Rhombus
All sides are equal

Rhomboid
Opposite sides are equal

Trapezoid
Two parallel sides

Trapezium
All sides and angles are different

Figure 4.8 Irregular four-sided polygons

Drawing a regular pentagon inside a circle

Figure 4.9 shows that to draw a regular pentagon inside a circle:

1 Draw diameter IJ.

2 Draw perpendicular OA to IJ where O is the centre of the circle.

3 Find point K, the mid-point of OJ.

4 Use a compass with arc radius KA and centre K to draw an arc that intersects with OI at point L. By geometry, it is found that AL equals the pentagon side.

5 Use a compass with arc radius AL to draw arcs crossing the circle perimeter at B, C, D and E, the pentagon corners.

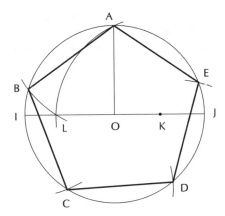

Figure 4.9 Drawing a regular pentagon inside a circle

Drawing a regular hexagon inside a circle

The characteristic of a regular hexagon, that its side is as long as the radius of the circle inside which the hexagon is drawn, is utilized. Using a compass with an arc radius equal to the circle radius, the hexagon corners can be located (see Fig. 4.10).

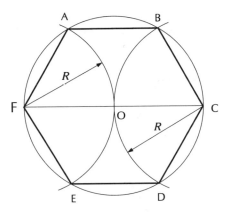

Figure 4.10 Drawing a regular hexagon inside a circle

Drawing a regular hexagon outside a circle

In this case, the angular characteristic of the regular hexagon is used to draw it, as shown in Fig. 4.11.

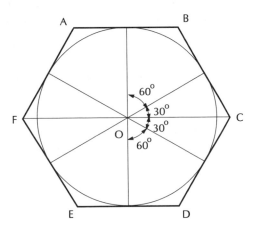

Figure 4.11 Drawing a regular hexagon outside a circle

Drawing a regular hexagon if its side length is known

A 30–60–90° triangle is used in this case, as shown in Fig. 4.12. Notice that FC is twice as long as the side length of the hexagon. Alternatively, a 30–60–90° triangle may be used to draw a regular hexagon, as shown in Fig. 4.13.

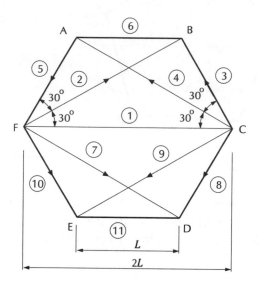

Figure 4.12   Drawing a regular hexagon, where 1 to 11 show the drawing steps

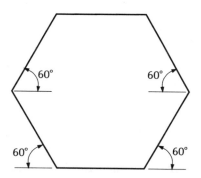

Figure 4.13   Drawing a regular hexagon

Drawing a regular octagon inside a square

To draw the required octagon, draw four arcs whose centres are at the square corners and whose radius is half the diameter of the square (see Fig. 4.14). The eight points that result from intersection of these four arcs with the square sides are the octagon corners.

Drawing a regular octagon inside a circle

From the centre of the circle draw lines with angles of 45° in between, as shown in Fig. 4.15. These lines cross the circle perimeter at the octagon corners.

Drawing a regular octagon outside a circle

This is the same as in the previous subsection, but the points of intersection now represent the points at which the octagon sides are tangent to the circle (see Fig. 4.16).

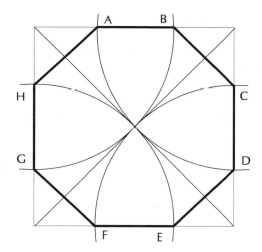

Figure 4.14 Drawing a regular octagon inside a square

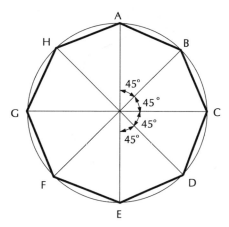

Figure 4.15 Drawing a regular octagon inside a circle

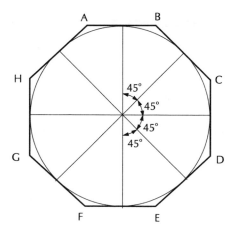

Figure 4.16 Drawing a regular octagon outside a circle

Drawing any regular polygon when the side length is known

As in Fig. 4.17, the following steps are carried out:

1 Draw a semi-circle whose centre is at A (one corner of the polygon) and radius is the polygon side length. Point B is the second corner.

2 Divide the perimeter of the semi-circle into a number of equal parts, equal to the number of polygon corners, by dividing angle 0AB.

3 Point G is a third corner of the polygon.

4 Draw lines joining A and points 3, 4, 5, etc., on the perimeter of the semi-circle.

5 Use a compass, with arc radius equal to the polygon side length, to draw an arc whose centre is G to cross line A3 at point F, the fourth corner.

6 Repeat step 5 to locate points E, D, etc., the polygon corners.

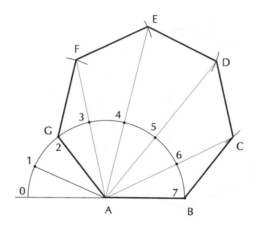

Figure 4.17   Drawing a regular polygon

4.5 Drawing tangents

In this section, drawing tangents to one or more circles is discussed.

Drawing a tangent to a circle from a point outside its perimeter

The steps to draw a tangent to circle O from point A are as illustrated in Fig. 4.18:

1 Draw a semi-circle whose diameter is OA to cross circle O at point B.

2 AB is a tangent to circle O.

Drawing a common tangent to two circles on the same side

The two circles shown in Fig. 4.19 have centres O_1 and O_2 and radii R_1 and R_2 respectively. To draw a common tangent on the same side, the following steps are carried out:

1 From O_1, the centre of the larger circle, draw a third circle whose radius equals $R_1 - R_2$.

2 Draw a semi-circle whose diameter is O_1O_2 to cross the third circle at point J.

3 Join O_1J and extend to point P_1 on the perimeter of circle O_1.

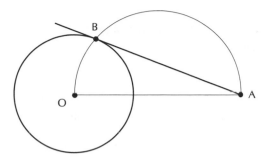

Figure 4.18 Drawing a tangent to a circle

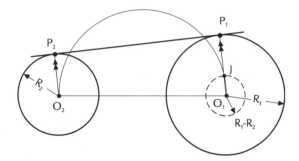

Figure 4.19 Drawing a common tangent to two circles

4   Draw line O_2P_2 parallel to O_1P_1 where P_2 is on the perimeter of circle O_2.

5   Line P_1P_2 is the common tangent to circles O_1 and O_2 on the outside.

Drawing a common tangent to two circles and crossing between them

For the two circles described in the previous subsection, a common tangent crossing between them is drawn by carrying out the following steps (see Fig. 4.20):

1   Draw perpendiculars O_1A and O_2B to O_1O_2.

2   Join AB and locate J, the intersection point of AB and O_1O_2. Notice that point J divides O_1O_2 such that $O_1J : O_2J = R_1 : R_2$.

3   Draw two circles whose diameters are O_1J and O_2J. These circles intersect with circles O_1 and O_2 at points P_1, P_2, Q_1 and Q_2.

4   Lines P_1Q_2 and P_2Q_1 are the common tangents to circles O_1 and O_2 on the inside.

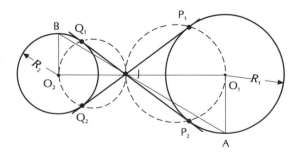

Figure 4.20 Drawing a common tangent to two circles

4.6 Drawing tangential arcs

Drawing an arc which is tangent to two lines or to a line and a circle is discussed in this section.

Drawing an arc tangent to two lines

Regardless of the angle between the two lines, the steps given below may be followed to draw an arc tangent to both lines (see Fig. 4.21).

1 Draw two parallel lines to the two given lines at a distance R away, where R is the arc radius.

2 The two additional lines intersect at the arc centre O.

3 Draw perpendiculars ON and OM to the two given lines, to locate the beginning and ending points of the arc.

4 From point O as a centre draw the required tangential arc using a compass.

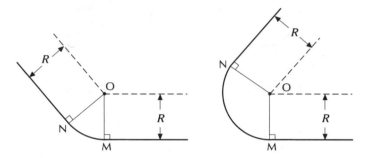

Figure 4.21   Drawing an arc tangent to two lines

Drawing an arc tangent to a line and a circle on the inside

For circle O and line AB, shown in Fig. 4.22, a common tangent arc on the inside is drawn in the following steps:

1 Draw an arc whose centre is O and radius is $R+R_1$, where R_1 is the radius of the required arc.

2 Draw line A_1B_1 parallel to AB at a distance R_1 away.

3 The drawn arc intersects line A_1B_1 at point O_1, the centre of the required arc.

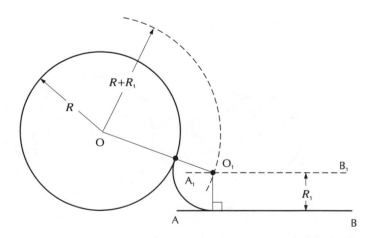

Figure 4.22   Drawing an arc tangent to a circle and a line

Drawing an arc tangent to a line and a circle on the outside

For the circle and line described in the previous subsection, a common tangent arc on the outside is drawn as follows (see Fig. 4.23):

1 Draw an arc whose centre is O and radius is $R_1 - R$, where R_1 is the radius of the tangent arc.

2 Draw line A_1B_1 parallel to line AB at a distance R_1 away.

3 The drawn arc intersects line A_1B_1 at point O_1, the centre of the required arc.

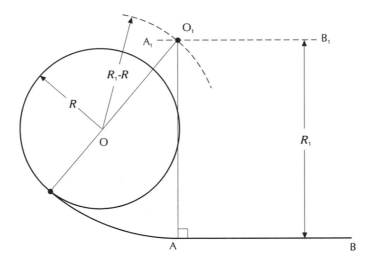

Figure 4.23 Drawing an arc tangent to a circle and a line

4.7 Conic sections

Conic sections are curved shapes resulting from the intersection of a plane and a cone. Three types of conic sections may result according to the angle between the plane and the longitudinal axis of the cone (see Fig. 4.24):

1 If the angle if larger than half the angle of the cone head, an ellipse results. A circle results if the angle equals 90°.

2 If the angle equals half the angle of the cone head, a parabola results.

3 If the angle is smaller than half the angle of the cone head, the intersection produces a hyperbola.

If the intersecting plane passes through the head point, two straight generators result (see Fig. 4.24).

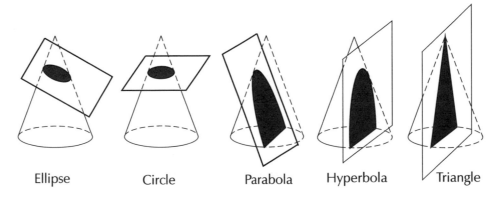

Ellipse Circle Parabola Hyperbola Triangle

Figure 4.24 Conic sections

4.8 Ellipses

An ellipse is a closed curved shape as shown in Fig. 4.24. Drawing an ellipse may be done using one of the following methods.

Drawing an ellipse by means of rays

1   Draw a rectangular envelope around the ellipse, or a rhomboid if the ellipse is in a plane inclined to the projection plane (refer to Fig. 4.25).

2   Divide the envelope short sides and the ellipse longest diameter into the same number of parts, eight in Fig. 4.25.

3   Draw rays from the ellipse top and bottom points towards the marked points on the envelope and the longest diameter.

4   Each two rays ending with the same number intersect at a point on the ellipse perimeter, e.g. rays D2 and B2 intersect at point P_2.

5   Step 4 results in a number of perimeter points sufficient to guide the drawing of the ellipse.

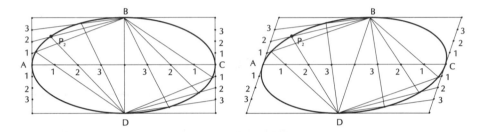

Figure 4.25   Drawing an ellipse by means of rays

Approximate drawing of an ellipse

This method depends on the characteristic of the ellipse (see Fig. 4.26), i.e. when a line crosses an ellipse, such that ob equals R_1 (half the longest diameter) and distance oa equals R_2 (half the shortest diameter). Based on this, an ellipse can be drawn as follows:

1   On a piece of paper with a straight edge, mark points o, a and b such that oa = R_2 and ob = R_1.

2   With points a and b moving along AB (the longest diameter) and CD (the shortest diameter) respectively, mark the positions of point o.

3   When enough points on the ellipse perimeter are marked, the ellipse is drawn.

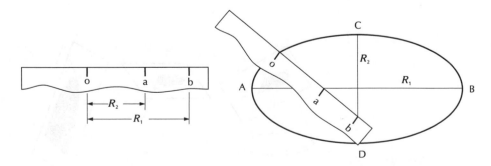

Figure 4.26   Approximate drawing of an ellipse

4.9 Parabolas

Drawing a parabola by means of rays

The parabola shown in Fig. 4.27 is drawn as follows:

1 Draw the rectangle that contains the parabola, or the rhomboid if the plane of the parabola is inclined to the projection plane.

2 Divide half the base and the height into an equal number of parts, four in Fig. 4.27.

3 Draw parallels to the centre line from the division points on the base.

4 Draw rays from head point H to side division points. The two sets of rays intersect at a series of points on the perimeter of the parabola.

5 Step 4 results in a number of points on the parabola perimeter, enough to draw it.

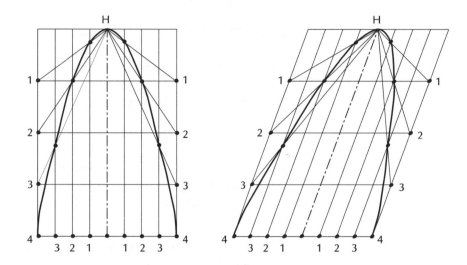

Figure 4.27 Drawing a parabola by means of rays

4.10 Hyperbolas

Drawing a hyperbola as a section of a cone

Figure 4.28 shows the construction of a hyperbola as a section of a cone carried out in the following steps:

1 Draw three views of a right circular cone and mark the plane that cuts the cone.

2 Draw an adequate number of cone generators that intersect with the plane and mark the points of intersection.

3 With enough points marked, the hyperbola is drawn as a smooth curve passing through these points.

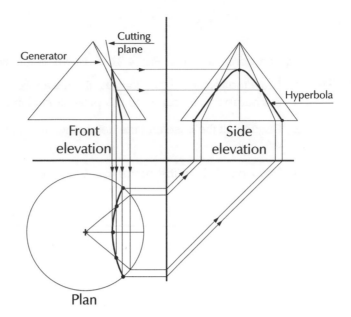

Figure 4.28   Drawing a hyperbola as a section of a cone

4.11 Construction of some special curves

Besides the conic sections, there are other mathematical curves that are frequently used in engineering and architectural work. The most common of these are the spiral of Archimedes, the helix and the cycloid.

The spiral of Archimedes

The spiral of Archimedes is a curve whose radius of curvature increases linearly with the angle through which it rotates. If R, the radius of the circle that contains the spiral after covering a 360° angle, is known, the spiral can be drawn as follows:

1   Divide the radius into a number of equal parts, say twelve as in Fig. 4.29, and draw a circle for each part.

2   Divide the 360° central angle into the same number of parts, and draw lines 01, 02, etc.

3   Mark the points at which the circles and lines of the same number intersect, e.g. circle 3 and line 03 intersect at point P3.

4   Draw the spiral as a smooth curve that passes through points P1, P2, P3, etc.

The helix

The helix is the three-dimensional curve which describes the path of a point that moves around a cylinder at a constant angular rate and also moves parallel to the axis at a constant linear rate. The helix is particularly important in the drawing of spiral staircases. Figure 4.30 shows the construction of a helix in the following steps:

1   Draw two views of the cylinder around which the point moves.

2   Divide the circular perimeter shown in plan into a number of equal parts, twelve in Fig. 4.30.

3   Mark the pitch of the helix (the rise that corresponds to one rotation) on the front elevation and divide it into the same number of equal parts.

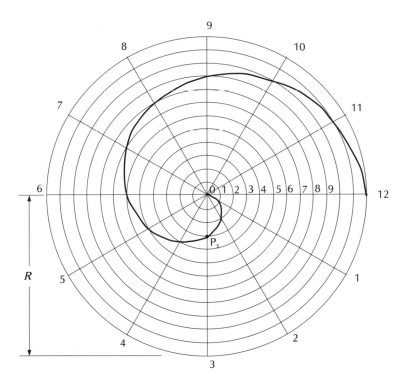

Figure 4.29 The spiral of Archimedes

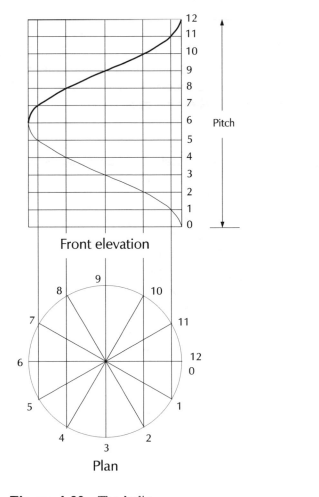

Figure 4.30 The helix

4   Draw vertical lines from points 1, 2, etc. on the plan to intersect with the pitch divisions on the front elevation at points on the helix.

5   Draw the helix as a smooth curve passing through these points.

The cycloid

The cycloid is the path of a point on the perimeter of a circle when the circle rolls on a horizontal flat surface. Figure 4.31 shows that, during rolling, the circle's centre O takes the positions O_1, O_2, etc., all on a horizontal line. The point under consideration starts at D_0 and ends at $D_{0'}$, a distance $2\pi R$ away. This distance is divided into twelve equal parts at points $D_{1'}$, $D_{2'}$, etc.

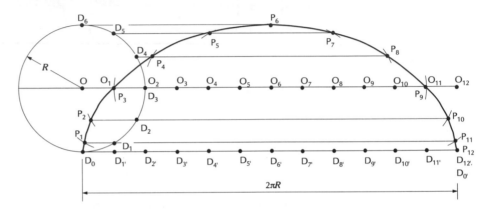

Figure 4.31   The cycloid

When the circle rolls until point D_1 coincides with point $D_{1'}$, point D_0 moves to point P_1 and the centre O moves to O_1. Point P_1 can be located such that it lies on the horizontal level of point D_1 and at a distance R from the new centre O_1. The same process is repeated to locate P_2, P_3, etc., through which a smooth curve is drawn to represent the cycloid.

4.12 Exercises

Draw to a proper scale the shapes shown in Figs 4.32 to 4.42.

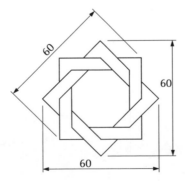

Figure 4.32

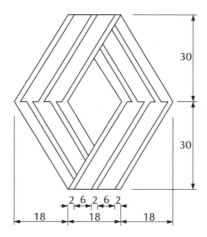

Figure 4.33

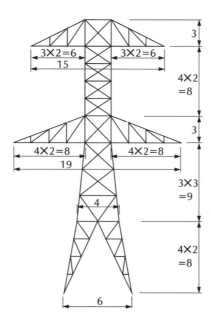

Figure 4.34

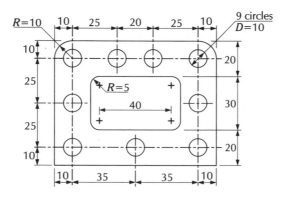

Figure 4.35

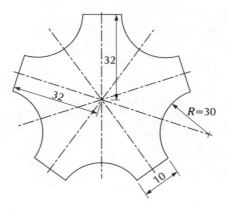

Figure 4.36

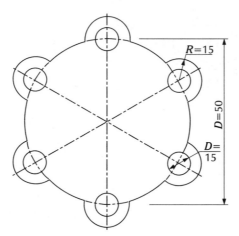

Figure 4.37

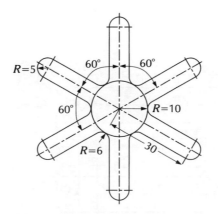

Figure 4.38

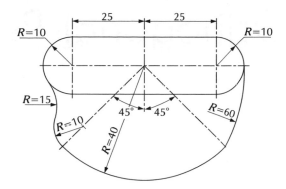

Figure 4.39

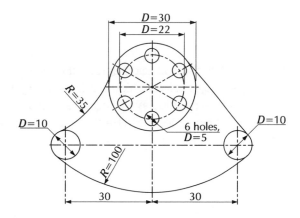

Figure 4.40

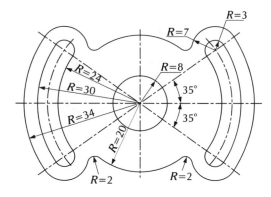

Figure 4.41

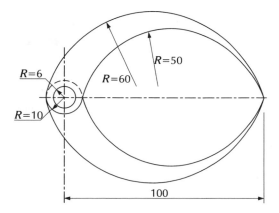

Figure 4.42

Orthogonal Projection

A projection is a drawing that represents a three-dimensional object on a two-dimensional surface. In engineering, the most common type of projection is the orthogonal projection in which the lines of sight from the eye to the object are parallel and perpendicular to the plane of projection (see Fig. 5.1). In orthogonal projection views, objects can be described precisely and fully with all the details given.

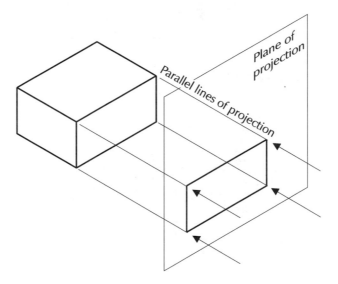

Figure 5.1   Orthogonal projection

5.1 Projection planes

The space in which the object exists can be divided by a horizontal plane and a vertical plane into four quadrants or angles I, II, III and IV, as shown in Fig. 5.2. The object can be assumed to exist in any angle (or quadrant), and three orthogonal views, a plan, a front elevation and a side elevation, are drawn accordingly. These views are respectively taken with respect to the horizontal plane, vertical plane and profile plane, also shown in Fig. 5.2.

The two types of orthogonal drawing commonly used in engineering and architectural drawings are first and third angle projection. Although drawings may be produced to either system according to BS1192, orthogonal drawings in this book are only drawn to first angle projection to reduce confusion. The symbols of first and third angle projection are shown in Fig. 5.3. These symbols

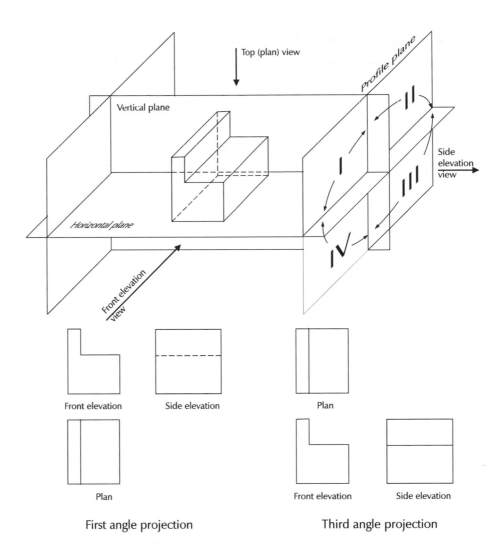

Figure 5.2 Orthogonal projection of an object

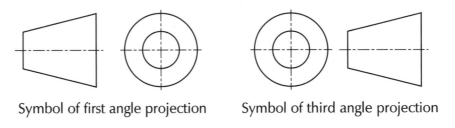

Symbol of first angle projection   Symbol of third angle projection

Figure 5.3 Symbols of first and third angle projection

should be presented on orthogonal projection drawings to specify the drawing system adopted.

It might be of interest to know that the first angle projection system is mainly a European system while the third angle system is North American. After the Second World War, the third angle system began to be more in use due to the interchange of drawings and technical information between Europe, on one hand, and the United States and Canada, on the other. However, neither system has any significant advantage over the other, and the reason for preferring one is merely a matter of tradition.

5.2 Projection of simple shapes and objects

Points are the main elements of any object, and the simple rules of projecting points on planes are utilized to carry out the projection of objects. For example, the projection of a straight line is a line connecting the projection of its two end-points.

Figure 5.4 shows the first angle projection of a horizontal rectangle. The plan view is identical to the rectangle as it is parallel to the horizontal plane of projection. In the front elevation, points A and D coincide at point A'' and line AD reduces to a point. Also in the side elevation, line AB reduces to point A'''. Figure 5.4b shows the three views as they should appear on a drawing sheet.

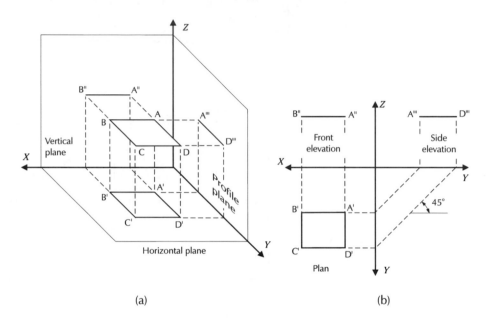

(a)   (b)

Figure 5.4   Orthogonal projection of a horizontal rectangle

Projection of three-dimensional objects is done in the same manner. The end-points of outlines are projected as before, and the projection of the object remains contained within the projection of its outlines.

The block shown in Fig. 5.5 has overall dimensions of $8 \times 6 \times 3$ in the X, Y and Z directions respectively. In each view, only one face is seen. The other parallel face has the same dimensions, and therefore disappears behind the front face. The four sides are reduced to the edges of the rectangle viewed.

5.3 Positioning of objects

Any object may be drawn efficiently in orthogonal projection. Theoretically, it may have any inclination to the planes of projection. However, as the purpose of orthogonal projection is to give a complete and accurate description of the object, it is recommended to consider the following guidelines:

1   The object is viewed so that its faces are parallel to the planes of projection (see Fig. 5.6).

2   The object is placed in its natural position or in the position in which it will be used. Never place a building with its height in the horizontal direction because, for example, the drawing space is too small.

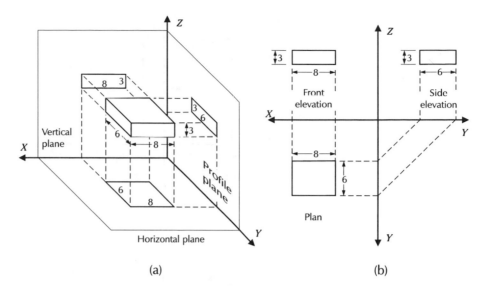

Figure 5.5 Orthogonal projection of a prism

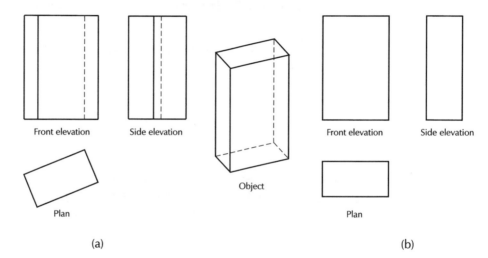

Figure 5.6 (a) Inconvenient and (b) convenient angles for orthogonal projection of a geometric object

3 If possible, the object is turned so that its most important face is parallel to the vertical plane.

5.4 Projection of a point on a surface

It is of vital importance for an engineer to be able to use two or three orthogonal views to visualize an object. The easiest way to learn this is to start with the projection of a point; then we move to the projection of geometric objects.

Projection of a point on a cylindrical surface

If the centre line of the cylinder is vertical, its plan view is a circle and the projection of any point on the cylinder surface is a point on the circle perimeter (refer to Fig. 5.7). In this case, the generator that passes through point P intersects the projection plane of the plan view at point P'. The horizontal distance, δ, between P' and generator 1 in plan equals the horizontal distance

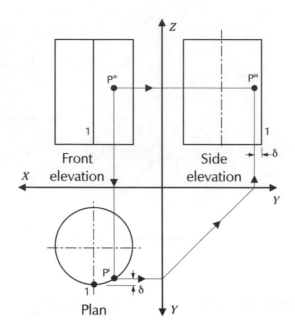

Figure 5.7 Projection of a point on a cylindrical surface

between generator 1 and the generator that passes through point P in the side elevation view.

If the centre line of the cylinder is horizontal, the same principles apply, although now the side elevation is a circle.

Projection of a point on a conical surface

The generator of the cone on which point P lies can be drawn on all three views as shown in Fig. 5.8. Point projection P″ is on the same horizontal level as P‴ and on the same generator which has been defined using the plan view.

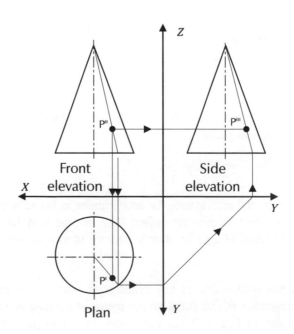

Figure 5.8 Projection of a point on a conical surface

Projection of a point on a prism

Figure 5.9 shows three views of a vertical prism with point P on one of its faces and point N on one of the edges. In plan, the projection of P is P' on the perimeter and under P'' on the front elevation. P''' on the side elevation is located as shown in the figure. The projection of point N is also shown in Fig. 5.9.

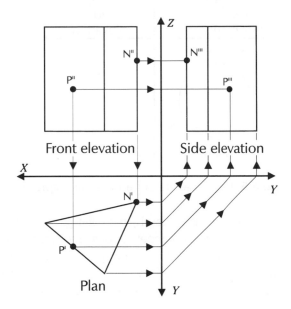

Figure 5.9 Projection of two points on a prism

Projection of a point on a pyramid

Figure 5.10 shows the projection of point P on a pyramid face and point N on one of its inclined edges.

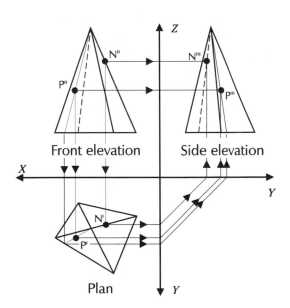

Figure 5.10 Projection of two points on a pyramid

Projection of a point on a sphere

The projection of point P on the surface of a sphere is shown in Fig. 5.11.

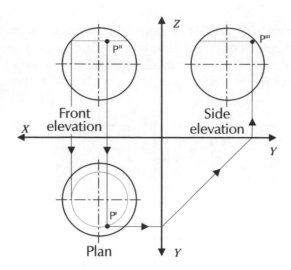

Figure 5.11 Projection of a point on a sphere

5.5 Orthogonal projection of objects

Geometric objects generally have flat or curved surfaces which meet at edges, and the edges, in turn, meet at points (corners). The easiest method to project an object is first to project its corner points, one by one. The projection of the object edges is the lines connecting the projected corner points, and the surfaces of the object are then readily projected, as each is surrounded by the projection of its edges.

If a surface is parallel to a projection plane, its projection on this plane is identical to its true shape. In each of the other two perpendicular planes, the projection is simply a straight line (see, for example, Fig. 5.4). On the other hand, an inclined surface to a projection plane is projected in a shape that depends on the angle of inclination.

The object shown in Fig. 5.12 has three faces, A, B and C, each of which is parallel to a projection plane. The three projections of each face are one that is identical to its true shape and the other two are straight lines. The inclined face D is projected differently.

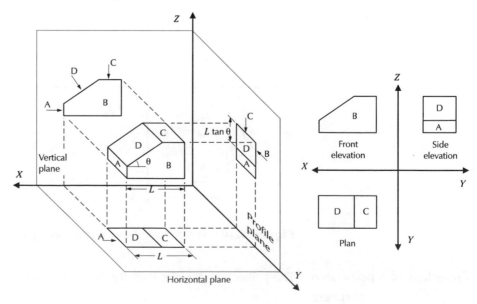

Figure 5.12 Projection of a geometric object

In the three orthogonal views shown in Fig. 5.13 of an object with a cylindrical part, the curved face 1–3–4 appeared in only the side elevation view because this face is parallel to the side vertical (profile) plane. In the other views, this face appeared as a line. Notice also that there is no line of separation between the projections of the cylindrical part and the neighbouring prism part as there is no sudden change of curvature between the surfaces of the two parts.

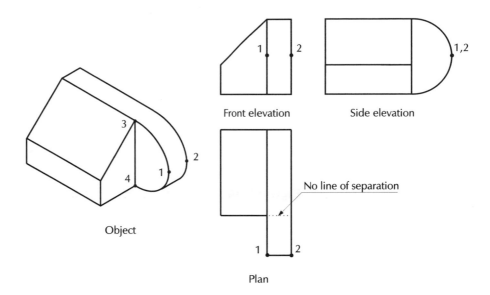

Figure 5.13 Projection of an object with a cylindrical part

Furthermore, notice that generator 1–2 is the edge line of the cylindrical part in the plan view as it is the last generator that can be viewed. In the front elevation view, this line does not appear as there is no associated change of curvature. The projection of generator 1–2 in the side elevation view is a point.

5.6 Projection of hidden lines

Most geometric bodies have edges and lines that do not appear from a certain direction. These lines are projected as dashed lines to distinguish them from other lines. (In some cases, hidden lines are ignored for clarity.) Figure 5.14 shows an object with some hidden lines and the projection of these as dashed lines in the three orthogonal views.

Figure 5.15 shows another object with two simple cylindrical holes, the projection of which in the elevation views is dashed lines. The centre lines of the cylindrical holes should be shown as well.

The elevation and plan views of plates with holes of different shapes and sizes are depicted in Fig. 5.16.

5.7 Two or three orthogonal views?

Most geometric bodies can be described fully by only two orthogonal projection views, mainly the plan and the front elevation or the front and side elevation. However, in some cases, all three views are required. For instance, the two elevation views shown in Fig. 5.17 may describe a cylinder or a prism. A plan is therefore needed in such a case.

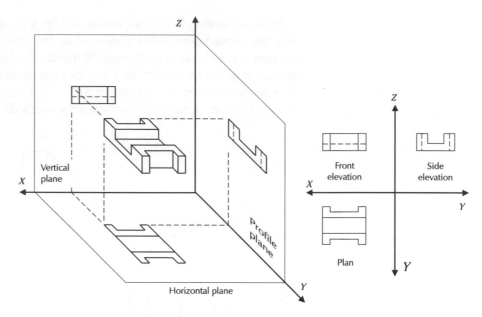

Figure 5.14 Projection of an object with hidden lines

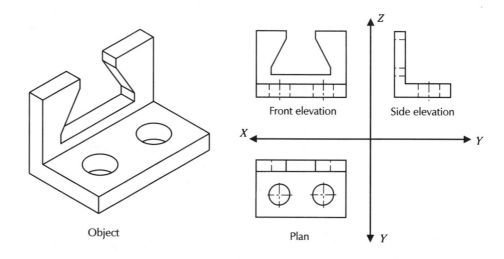

Figure 5.15 Projection of an object with cylindrical holes

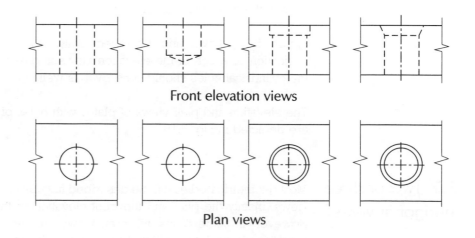

Figure 5.16 Projection of plates with cylindrical holes

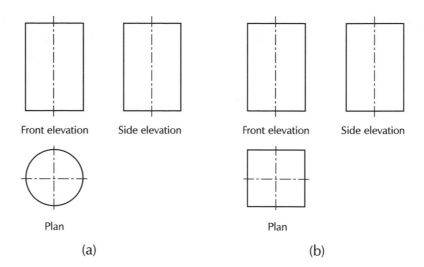

Figure 5.17 The similarity between (a) the projection of a cylinder and (b) that of a prism

Another example is shown in Fig. 5.18 where, for the given plan and front elevation views, two possible side elevation views are presented leading to two completely different interpretations of the drawings.

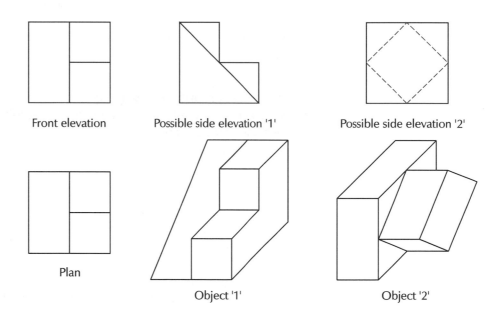

Figure 5.18 Two different interpretations of the given plan and front elevation views

In orthogonal projection, the front elevation is known to be the most important among all views since it is easier to visualize an object for which a front elevation and a plan, or a front and side elevation, are drawn than with a plan and a side elevation. Since the latter two views are on different levels on the drawing sheet, it is difficult for the reader to develop a visual image of the object.

5.8 Sectional views

For objects with internal as well as external profiles, the use of dashed lines to describe the hidden profiles may prove confusing. In such cases, it is common to imagine that a part of the object is cut out in a way suitable to expose the hidden profiles. The exposed parts can then be drawn as full lines. Such a view is known as a section, and the cutting plane the section plane. Figure 5.19 shows an object cut by a section plane AA to expose its internal details.

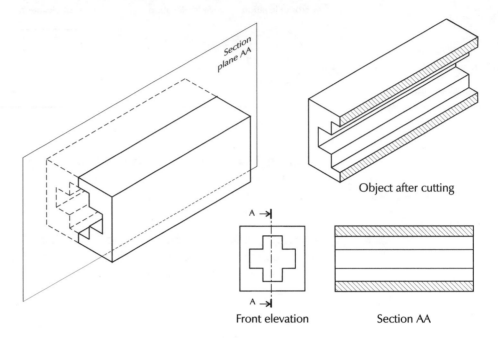

Figure 5.19   Exposing the internal profile of an object by means of a section plane

A further example is that of a single-storey house, shown in Fig. 5.20. The section views taken can adequately be used to describe the internal details and dimensions of the house.

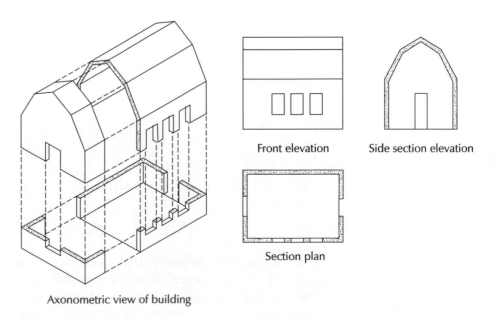

Figure 5.20   Exposing the internal details of a house

Section markers are used to locate the section plane and the direction of viewing. Section markers must be clear and simple. Some of the common styles of section markers are shown in Fig. 5.21.

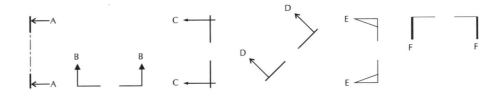

Figure 5.21   Common styles of selection markers

The section plane must be indicated on at least one orthogonal view and lettered for easy reference. The section letters must be clear and readable from the bottom of the drawing. The material cut by the section plane is cross-hatched by thin diagonal lines, normally at 45° to the horizontal for metals. For other materials see Fig. 5.22.

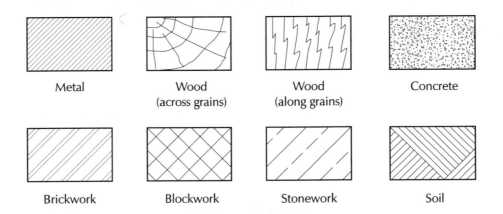

Figure 5.22   Sections through different materials

When several components of an object are cut by the same section plane, each component must be hatched differently so that it can be identified easily. For this purpose, the hatching lines may have different spacings and the 45° angle may be reversed (see Fig. 5.23). The angle of hatching may also be changed to, say, 30 or 60°.

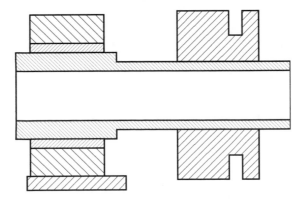

Figure 5.23   Section through an object made from several components

When sections are taken through objects with inclined edges, the 45° hatching lines might not be appropriate and lines with a different inclination may be used (see Fig. 5.24a). Also, when section planes cut through parts of objects that are too thin to draw in hatching lines, these parts may be blackened as in Fig. 5.24b. However, when two such parts are adjacent in a section, they should be separated by a thin space.

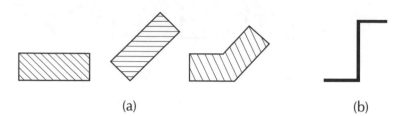

(a)   (b)

Figure 5.24   Section hatching

When a section plane cuts a large area, hatching may be applied only at the area borders while the middle parts are left blank (see Fig. 5.25a). Also, if a long area is cut, hatching may be applied only the ends, as shown in Fig. 5.25b.

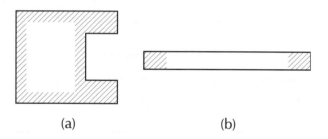

(a)   (b)

Figure 5.25   Hatching of (a) a large area and (b) a long area

For cases when a dimension figure must be written inside a cut part, the section hatching lines must be broken to avoid cutting across the figures, as depicted in Fig. 5.26.

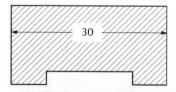

Figure 5.26   Breaking the section hatching lines to include a dimension figure

If a component is projected (while cut) in more than one sectional view, the spacing and angle of hatching lines of this component should remain the same in all sectional views (see Fig. 5.27).

Half sections   When a symmetrical object has internal details, it may be appropriate to draw a half-sectional view in which the internal details are exposed in one half of the view while the other half remains a normal view (see Fig. 5.28).

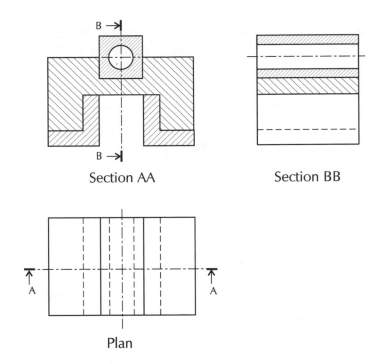

Section AA   Section BB

Plan

Figure 5.27 Matching the spacing and angle of hatching lines

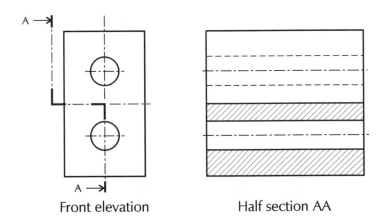

Front elevation   Half section AA

Figure 5.28 Projection of a symmetrical object with internal details

Part sections   When an object has only a local internal detail, it is commonly sufficient to expose this part by means of a part section (see Fig. 5.29).

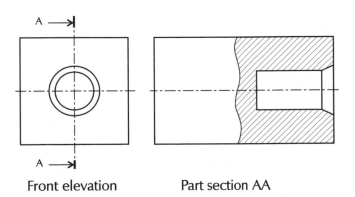

Front elevation   Part section AA

Figure 5.29 Projection of an object with a local internal detail

Staggered sections   For an object which has internal details that are not on one line, a staggered section may be appropriate (see Fig. 5.30).

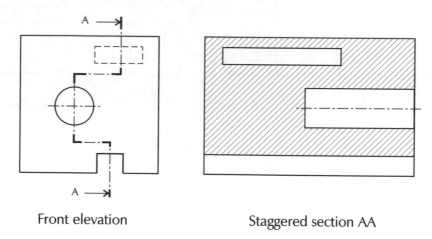

Front elevation   Staggered section AA

Figure 5.30   Projection of an object with staggered internal details

Revolved sections   When a revolved section of one simple part of an object is required, it may be drawn directly on the part under consideration. As an example, Fig. 5.31 shows the details of a reinforced concrete beam/slab floor.

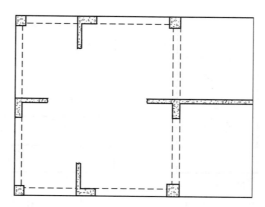

Figure 5.31   Revolved sections of beams and slabs drawn on a plan of a concrete floor

5.9 Dimensioning of orthogonal drawings
Besides the general rules for the dimensioning of geometric bodies discussed in Sec. 3.2, additional rules related to civil engineering applications are introduced in this section.

Levels   Levels record the vertical distance of a position above or below a datum. On a plan view, the level of a position is indicated by giving the level magnitude adjacent to the position, which may be located by a vertical cross. Levels on vertical views are presented as shown in Fig. 5.32.

Grids   Grids may be used to dimension building components in layout views. In most grids, two perpendicular sets of axes are used. While numerals are used to define the axes in one direction, letters define the axes in the other direction,

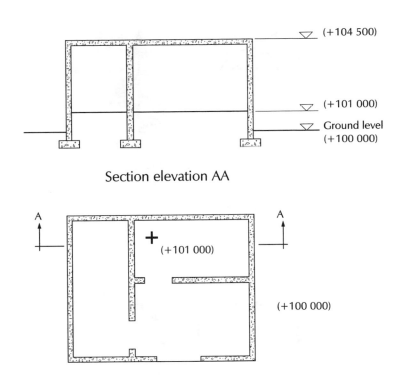

Section elevation AA

Section plan

Figure 5.32 Levels on plan and elevation views

with I and O commonly omitted (see Fig. 5.33). On such layout drawings, the load-bearing components of a building, beams and slabs of a floor, etc., may be presented. As an example the columns shown in the plan view of Fig. 5.33 can be easily located using the rectangular grid presented.

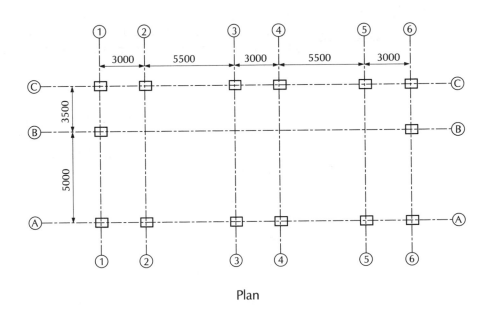

Plan

Figure 5.33 Columns on a rectangular grid

North sign On layout drawings, the direction of north should be clearly specified. A symbol similar to one of those shown in Fig. 5.34 may be used.

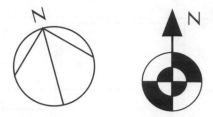

Figure 5.34 North signs

Holes   When a section is taken through a hole and the surrounding solid material is adequately hatched, the hole becomes evident and requires no further clarification. However, when an outside view is taken, holes are presented edge-shaded as shown in Fig. 5.35.

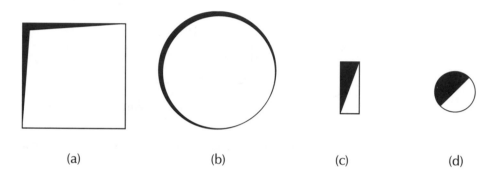

(a)   (b)   (c)   (d)

Figure 5.35 Holes: (a) large rectangular, (b) large circular, (c) small rectangular, (d) small circular

5.10 Drawing procedure

Engineering and architectural drawings must be accurate and neat. The industrial requirements demand additionally that they should be done within acceptable time limits. To satisfy these requirements and to help the draughtsmen draw efficiently, it is recommended that the following order of procedure be considered (refer to Fig. 5.36):

(a) Determine the sizes of the views according to the scale chosen. Their overall layout is then drawn while leaving proper spacing in between. Starting with the outlines facilitates detecting errors such as scale and misfitting errors.

(b) Main centre lines and datum lines are then drawn.

(c) It is better to draw parts and components in all views simultaneously than finishing the views one by one.

(d) After drawing the outlines and centre lines, draw the curves with finished lines before drawing any other finished straight lines. Straight lines that are tangential to the curves may then be finished while blending neatly with the finished curves.

(e) Unwanted working and construction lines are erased before final completion of lining.

(f) Dimensions, instruction notes, titles and scales are finally inserted.

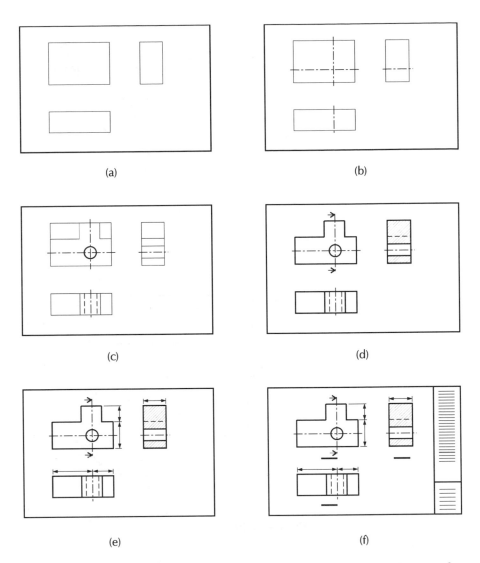

Figure 5.36 Recommended order of drawing procedure: (a) layout of drawing, (b) centre lines, (c) draw views simultaneously, (d) clean and finish lining, (e) add dimensions, (f) titles and notes

5.11 Exercises

1 Draw three orthogonal views for each of the objects shown in Figs 5.37 to 5.54.

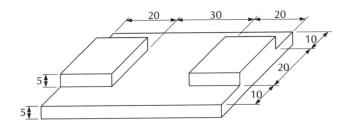

Figure 5.37

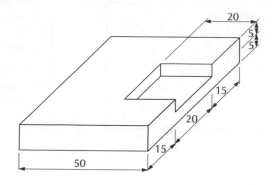

Figure 5.38

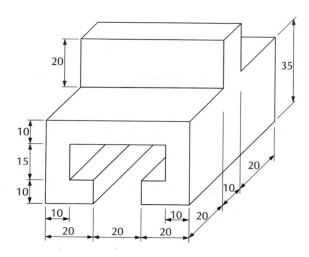

Figure 5.39

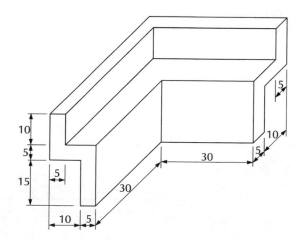

Figure 5.40

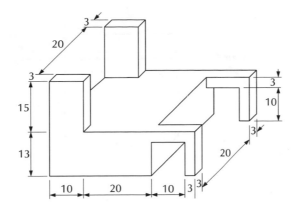

Figure 5.41

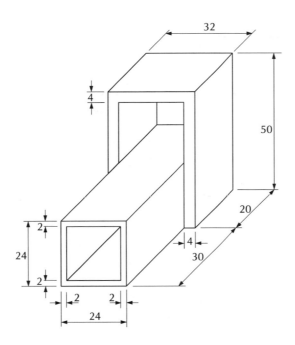

Figure 5.42

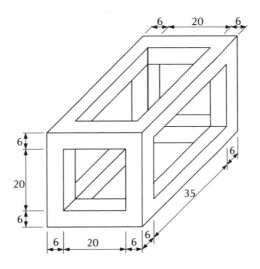

Figure 5.43

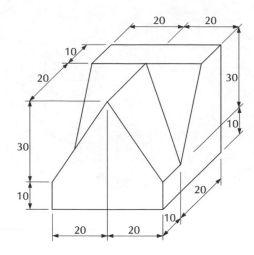

Figure 5.44

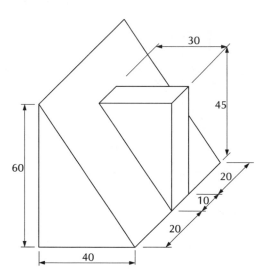

Figure 5.45

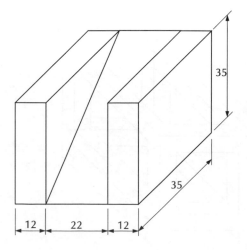

Figure 5.46

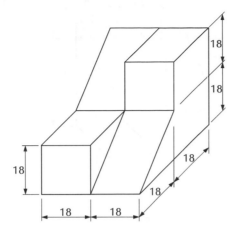

Figure 5.47

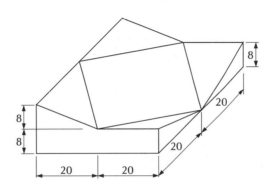

Figure 5.48

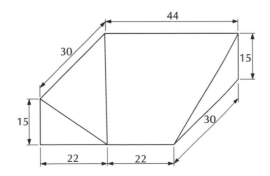

Figure 5.49

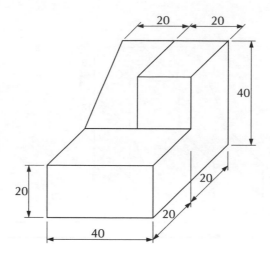

Figure 5.50

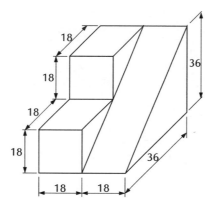

Figure 5.51

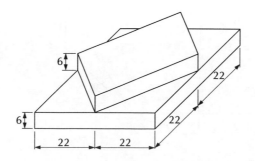

Figure 5.52

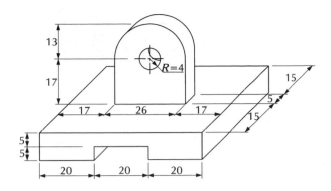

Figure 5.53

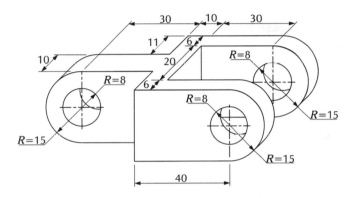

Figure 5.54

2 The objects shown in Figs 5.55 to 5.58 present well-known civil engineering structures. Draw three orthogonal views for each object to a proper scale.

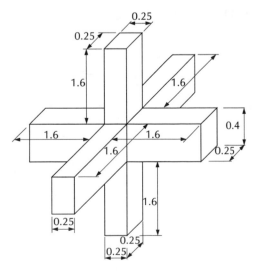

Figure 5.55 Beam/column connection

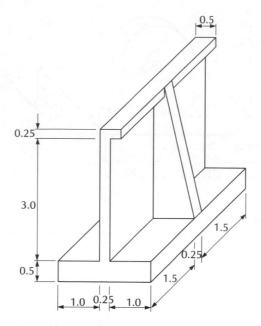

Figure 5.56 Retaining wall

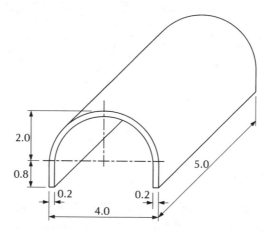

Figure 5.57 Cylindrical shell

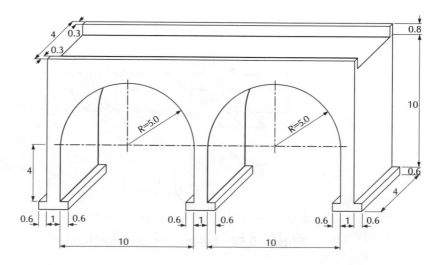

Figure 5.58 Bridge

3 For the objects presented in Figs 5.59 to 5.63 draw the required sectional view.

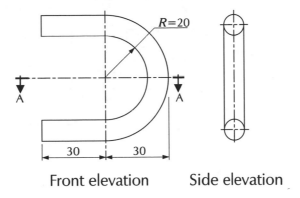

Front elevation   Side elevation

Figure 5.59

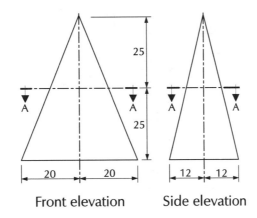

Front elevation   Side elevation

Figure 5.60

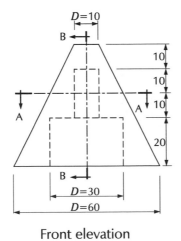

Front elevation

Figure 5.61

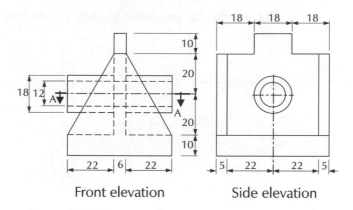

Front elevation   Side elevation

Figure 5.62

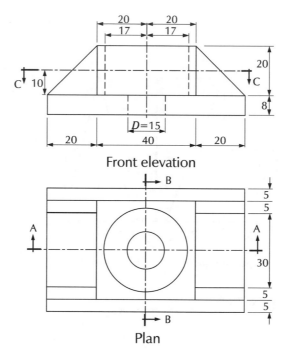

Front elevation

Plan

Figure 5.63

4   Draw a third orthogonal view for each of the objects shown in Figs 5.64 to 5.73.

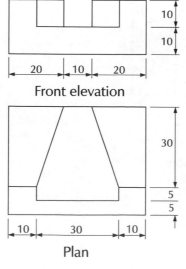

Front elevation

Plan

Figure 5.64

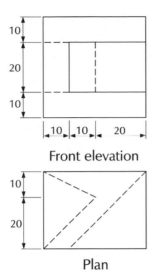

10
20
10
10
10
20
Front elevation
10
20
Plan

Figure 5.65

20
10
10
20
20
20
10
10
Front elevation
Side elevation

Figure 5.66

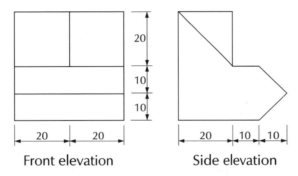

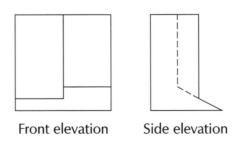

Front elevation
Side elevation

Figure 5.67

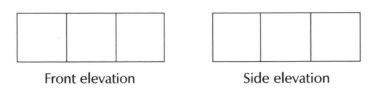

Front elevation
Side elevation

Figure 5.68

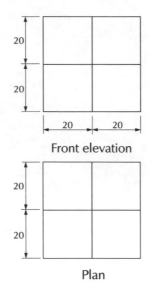

Front elevation

Plan

Figure 5.69

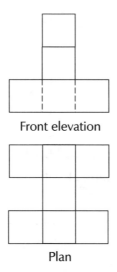

Front elevation

Plan

Figure 5.70

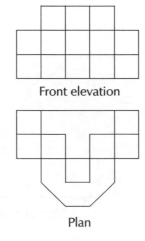

Front elevation

Plan

Figure 5.71

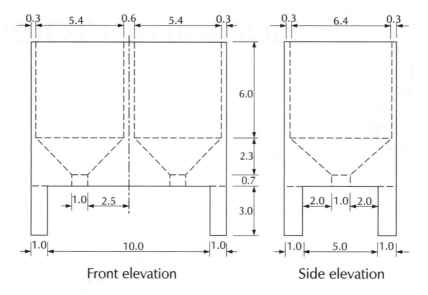

Front elevation   Side elevation

Figure 5.72

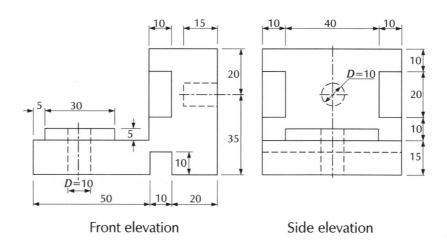

Front elevation   Side elevation

Figure 5.73

CHAPTER 6

Isometric Projection

Isometric projection is a type of axonometric projection which in turn is a division of pictorial projection. Isometric and oblique projections (another type of pictorial projection, discussed in Chapter 7) are the most commonly used techniques of pictorial projection in civil engineering applications. This chapter begins with an introduction to pictorial projection in general and axonometric projection in particular, before introducing isometric projection.

6.1 Introduction to pictorial projection

Pictorial projection is a general term that refers to all projections which create a three-dimensional impression of objects by depicting length, width and height on one view. The pictorial view is easily visualized by even inexperienced readers. However, it has the drawback of being incomplete with the need to overlook some details for reasons of clarity.

The three main classifications of pictorial projection are the axonometric, the oblique and the perspective projections. Each group can be further subdivided, and has its own rules and uses (see Fig. 6.1).

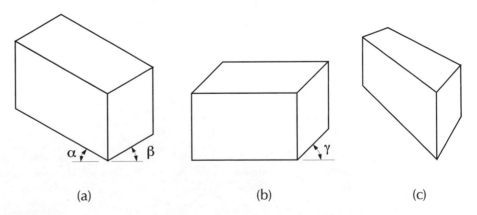

(a)   (b)   (c)

Figure 6.1   Types of pictorial projection: (a) axonometric, (b) oblique, (c) perspective

Axonometric projection

In the axonometric type of projection, the projector lines from the object to the picture plane are perpendicular to this plane and parallel to each other. Unlike orthogonal projection, in axonometric projection the picture plane is inclined to all main horizontal and vertical planes of the space. The special type of axonometric projection where the angles of inclination to all three planes are equal is called isometric projection (see Fig 6.2a). In this case:

1 The three main axes X, Y and $Z^\dagger$ are projected on a drawing sheet so that they are at 120° to each other.

2 All three dimensions, length, width and height, are projected with the same scale. Commonly, a scale of 1:1 is used for simplicity.

3 Parallel lines on the actual object are projected parallel on the view.

Other types of axonometric projection are in use, they include:

1 The dimetric projection, where the object is set so that two of its three dimensions are equally foreshortened on the picture plane. The third dimension is foreshortened to a different degree. Two different scales must be used in this case (see Fig. 6.2b).

2 The trimetric projection, where the three dimensions of the object are foreshortened to three different scales, as shown in Fig. 6.2c.

The isometric type of axonometric projection is commonly preferred by engineers to other types due to its simplicity and ease of use. For this reason, only the isometric projection is focused on in this chapter.

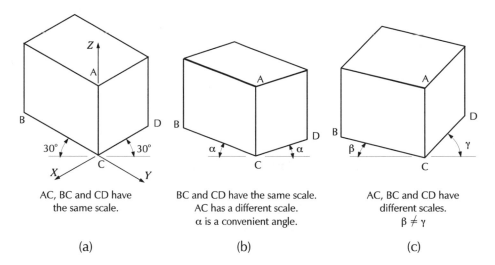

AC, BC and CD have the same scale.

BC and CD have the same scale.
AC has a different scale.
α is a convenient angle.

AC, BC and CD have different scales.
$\beta \neq \gamma$

(a)   (b)   (c)

Figure 6.2 Types of axonometric projection: (a) isometric, (b) dimetric, (c) trimetric

6.2 Isometric projection of lines

In isometric views, lines that are parallel to any of the three main axes X, Y and Z are projected with their true lengths and called isometric lines. However, for lines of other inclinations the projected length does not match the true length. These can only be projected by locating the projection of their end-points. Figure 6.3 shows a geometric object with some inclined edges and how these are projected isometrically.

---

† The main axes X, Y and Z are the intersection lines of the three main projection planes.

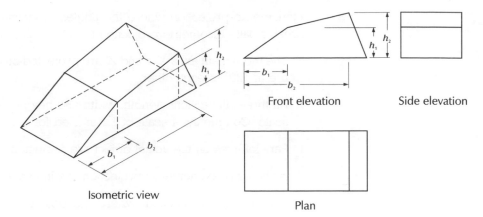

Figure 6.3   Isometric projection of an object with inclined edges

6.3 Isometric projection of angles

Angles can not be drawn with their true value in isometric views. The coordinates of points on the two lines that contain an angle may be used to project it. Alternatively, the tangent value of the angle may be used (see Fig. 6.4).

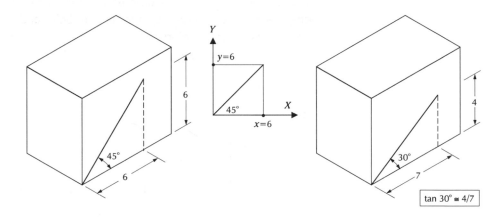

Figure 6.4   Isometric projection of angles

6.4 Isometric projection of circles and curves

The isometric projection of a circle is an ellipse, and this can be drawn once the circle's centre, radius and plane are determined. The first step in drawing the projection of a circle is always determining the plane in which it lies. There are two methods to draw a circle when it is parallel to any of the main projection planes. Otherwise, a general method mainly developed to project general curves is used.

Method of four centres (approximate method)

1   Draw rhombus ABCD by drawing lines AB and DC parallel to diameter HF and lines BC and AD parallel to diameter EG. This rhombus is an envelope to the circle projection (see Fig. 6.5).

2   Obtuse angle heads B and D are used as centres to draw arcs HG and EF. Notice that line BH is perpendicular to AD, etc.; therefore arcs HG and EF are tangents to the rhombus sides.

3   Draw lines BH and DE and lines BG and DF. The two pairs intersect at points I and J respectively.

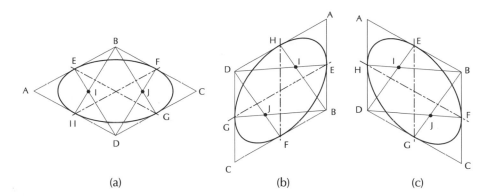

Figure 6.5 Isometric projection of circles by the method of four centres: (a) *XY* plane, (b) *ZX* plane, (c) *YZ* plane

4 From points I and J as centres, draw arcs EH and FG. Since IH is perpendicular to AD, etc., arcs EH and GF are also tangents to the rhombus sides.

Method of intersecting lines

The isometric drawing of a circle may be done by following the steps given below (refer to Fig. 6.6):

1 Draw the circle with its true size (Fig. 6.6a).

2 Draw a series of horizontal and vertical lines that intersect with the circle perimeter and mark the points of intersection, 1, 2, etc.

3 Project these lines on the isometric view and locate all the previously marked points of intersection, 1′, 2′, etc. These points are finally connected by a smooth curve to represent the projection of the circle.

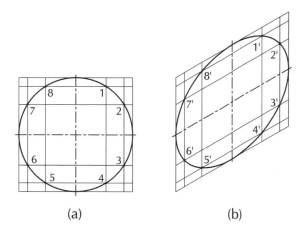

Figure 6.6 Isometric projection of a circle by the method of intersecting lines

General method of projecting curves

The method illustrated in this section is general for any curve (or a circle) whatever the plane in which the curve lies. In general, any curve is a set of lines connecting a group of points. Projection of these points on an isometric view is done using the coordinates of each point. The projection of the curve is finally done by connecting the projected points by a smooth curve (see Fig. 6.7).

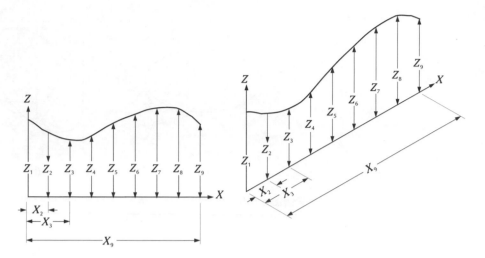

Figure 6.7   Isometric projection of general curves

Common errors in drawing circles

Two common errors are frequently made by students and draughtsmen:

1   Circles may be drawn out of the proper isometric plane. Notice that the sides of the envelop, in which the circle is drawn, must be parallel to the two main axes of the plane in which the circle lies (see Fig. 6.8a).

2   When short cylinders or cylindrical holes are drawn, the tangent between the end circles is frequently omitted. Also, the rear ellipse that represents the far-end circle is sometimes missing (see Fig. 6.8b).

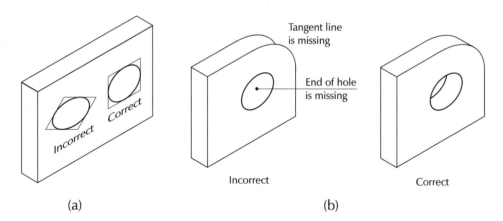

Figure 6.8   Common errors in the isometric projection of circles

6.5 Isometric projection of three-dimensional objects

Moving from drawing plane figures to drawing three-dimensional objects is a simple step, involving only the consideration of the third dimension. Isometric projection of an object is commonly done through the following steps:

1   Draw the orthogonal views of the object and enclose them in the smallest rectangular box possible. The scale used is the one to be adopted in the required isometric view.

2   Draw the enclosing box in isometric with the edges at 120° to each other.

3 Start with the parts that lie at, or adjacent to, the faces of the box and then proceed away from there. Points are first projected using their coordinates; then lines are drawn to form the edges of faces and surfaces.

4 It is a common practice to omit all hidden parts and dashed lines for clarity.

As an example, refer to Fig. 6.9, which shows the steps of drawing an isometric view of an object.

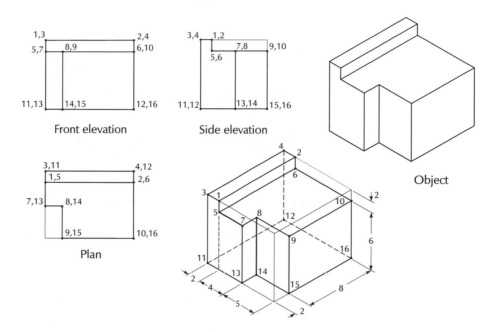

Figure 6.9 Isometric projection of an object

6.6 Disadvantages of isometric projection

Isometric projection, although easy and quick to use, suffers from some disadvantages, the first of which is general to all pictorial projection techniques, that is drawings are not complete with some information missing. What is more serious is that objects appear in isometric views to be larger than they truly are. For example, the isometric projection of a circle is an ellipse whose larger diameter is longer than the true dimension. This might be argued to be insignificant as the user of isometric views is always reminded to consider only the dimensions in the directions of the three main axes X, Y and Z. Dimensions in all other directions are not true. However, the case of a sphere is a peculiar one, as it should appear with the same size regardless of the inclination of the projection plane. A wrong impression will surely be gained if the sphere is drawn with its true diameter while other objects are enlarged. A common solution to this particular case is to project the sphere while considering enlarged dimensions. This results in an apparently larger sphere, but remember that diameters in all three X, Y and Z directions still equal the true diameter.

In brief, the user of isometric projection must be aware of its shortfalls as much as its merits. With this in mind, drawings that are suitable for the job in hand could be produced and realistic interpretation of existing drawings could be developed. The user must also learn to distinguish between the directions of the true dimensions and the overall wrong impression created by the view.

6.7 Dimensioning of isometric drawings

Besides the requirements related to orthogonal drawing dimensioning and discussed in Sec. 3.2, additional rules are introduced in this section for isometric drawing:

1 All dimension and extension lines must be isometric lines and lie in isometric planes. A common error is illustrated in Fig. 6.10 where the dimension and extension lines do not lie in one isometric plane, although the dimension line is vertical.

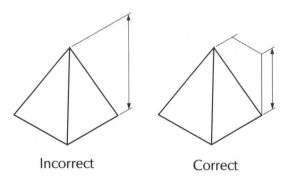

Incorrect   Correct

Figure 6.10 A common error in dimensioning isometric views

2 All dimension figures must lie in isometric planes to which they are related; otherwise vertical writing of figures is also acceptable (see Fig. 6.11).

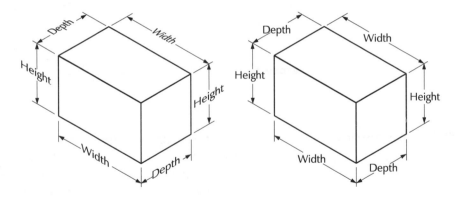

Figure 6.11 Possible systems of dimensioning isometric views

Figure 6.12 illustrates the dimensioning of a geometric object according to the two systems discussed above.

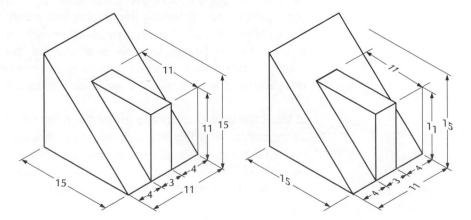

Figure 6.12 Dimensioning of a geometric object

6.8 Exercises Draw isometric views for the objects shown in Figs 6.13 to 6.24.

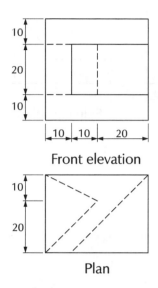

Front elevation

Plan

Figure 6.13

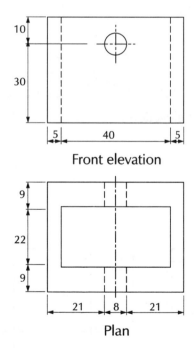

Front elevation

Plan

Figure 6.14

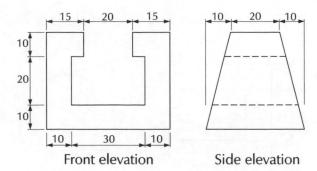

Front elevation   Side elevation

Figure 6.15

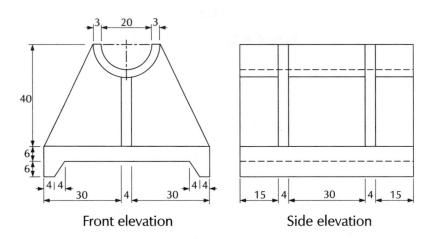

Front elevation   Side elevation

Figure 6.16

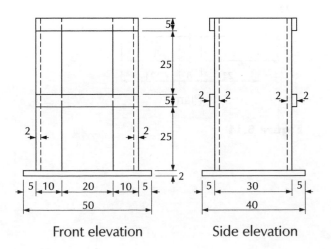

Front elevation   Side elevation

Figure 6.17

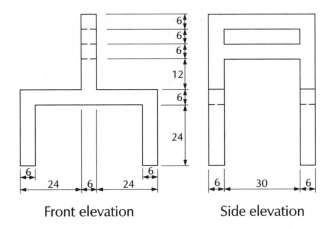

Front elevation   Side elevation

Figure 6.18

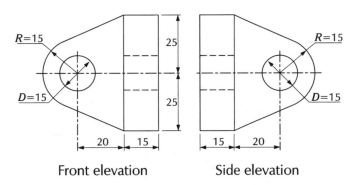

Front elevation   Side elevation

Figure 6.19

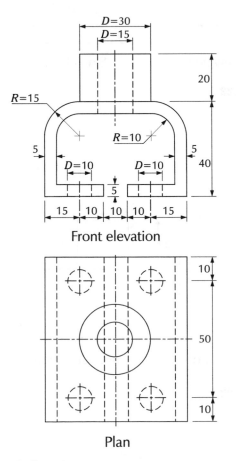

Front elevation

Plan

Figure 6.20

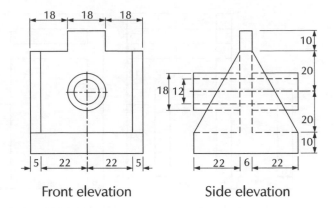

Front elevation   Side elevation

Figure 6.21

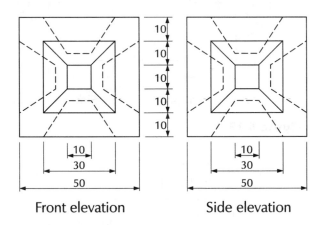

Front elevation   Side elevation

Figure 6.22

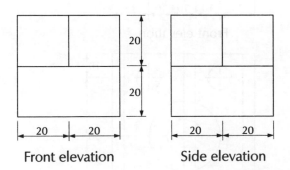

Front elevation   Side elevation

Figure 6.23

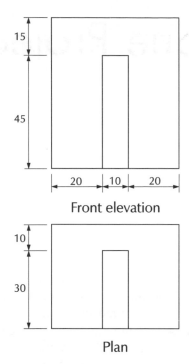

Front elevation

Plan

Figure 6.24

Oblique Projection

7.1 Introduction

Besides isometric drawing, oblique projection is very popular in civil engineering and architectural applications. The following sections describe the characteristics of this technique of drawing and how it is best used. Drawn examples are presented throughout the chapter to illustrate the technique. Finally, there is a discussion on the disadvantages of oblique projection and how it compares with isometric drawing. While oblique views do not involve an alteration to the object size, as isometric views do, they sometimes result in an undesirable degree of face distortion.

Similar to orthogonal and axonometric projections, the projection lines in oblique views are parallel to each other, but in the present case they are oblique to the plane of projection. Figure 7.1 illustrates this fact.

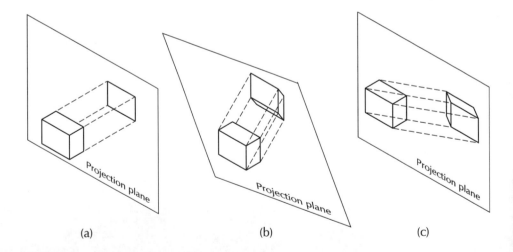

(a)   (b)   (c)

Figure 7.1 Projection lines in the three projection techniques: (a) orthogonal, (b) axonometric, (c) oblique

Since in oblique views projection lines are parallel to each other, any line that is parallel to the projection plane will project with its true length and parallel to its original position. Consequently, any parallel face to the projection plane retains its true shape and size, and its projection is identical to its orthogonal view. This in fact features one of the main advantages of oblique projection.

Figure 7.2a depicts a typical oblique view of a prismatic object where the front face, and all parallel faces, retain their true shape and size. However, in some cases it might be more appropriate to draw oblique views as depicted in Fig. 7.2b, with the top face, and all parallel faces, retaining their true shapes and dimensions. This might be useful in some civil engineering and architectural applications, for example when a designer removes the roof of a building to show its internal details (see Fig. 7.2c). These views are easy to produce since the top face is identical to the plan orthogonal view after being rotated 45°. The discussion in this chapter is limited to the first technique illustrated in Fig. 7.2a since it is the most commonly used in practice. However, the same principles apply to the second technique without alteration.

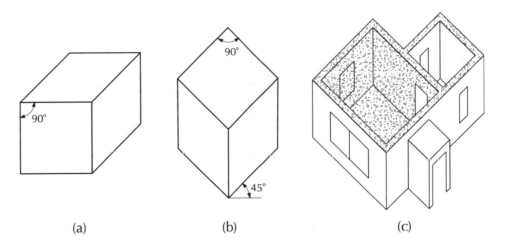

(a) (b) (c)

Figure 7.2 (a) Front face retains shape and size. (b) Top face retains shape and size. (c) Oblique view of a building

In oblique views, the angle between the projection lines and the projection plane determines the scale at which a line perpendicular to the projection plane is foreshortened. This angle may be varied as desired, but for practical purposes only two values of 45° and 63° 26′ are adopted in the most important types of oblique projection: Cavalier and Cabinet projection respectively.

Cavalier projection   In Cavalier projection, projection lines make an angle of 45° with the projection plane. In this case, all X, Y and Z oriented lines of an object are projected in their true lengths (see Fig. 7.3 for a Cavalier projection of a cube of 80 mm side length). This features a major advantage of Cavalier views over any other form of oblique projection. For this reason, Cavalier projection is the most popular form of oblique projection in civil engineering and architectural applications, and therefore attention is focused on it in the remainder of this chapter.

On the other hand, Cavalier projection, like all forms of oblique projection, suffers from the undesirable feature of distortion of the side and top faces. This will be discussed further in the following sections.

Cabinet projection   In Cabinet projection, the lines of sight make an angle of 63° 26′ with the projection plane. While faces that are parallel to the projection plane are projected in their true size, perpendicular lines are projected with only half of their length. As an example, a Cabinet projection of a cube of 80 mm side length is as shown in Fig. 7.4.

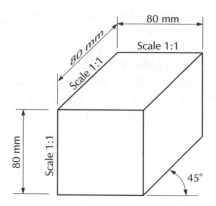

Figure 7.3   Cavalier projection of a cube of 80 mm side length

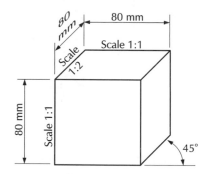

Figure 7.4   Cabinet projection of a cube of 80 mm side length

7.2 Positioning of objects

Any object may be drawn efficiently in oblique projection with, theoretically speaking, any position relative to the projection plane. However, to obtain the best oblique view possible, it is recommended that the following guidelines be considered:

1 Since the front face of the object retains its true shape and size, it is best to position the object so that its front face is the most important face (refer to Fig. 7.5).

2 If the object contains circular parts, it is advisable to position the object so that its circular parts are on the front face, or on faces parallel to the front face (see Fig. 7.6).

3 Distortion is one of the disadvantages of oblique projection. To reduce its effect on the quality of oblique views it is recommended that objects be positioned so that their largest dimension is parallel to the projection plane (see, for example, Fig. 7.7).

4 In addition to recommendations 1 to 3 above, it is advisable to position the object so that as few parts as possible are hidden in the oblique view. As an example, position (d) of Fig. 7.7 with no hidden parts is better than position (b). Figure 7.8 shows another example where the object was rotated 90° to avoid having hidden parts in the oblique view.

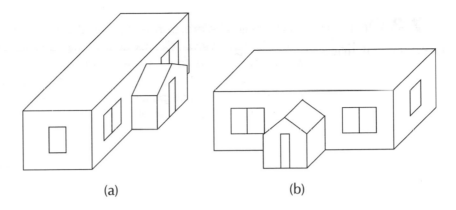

(a)   (b)

Figure 7.5 (a) Inconvenient and (b) convenient object positions

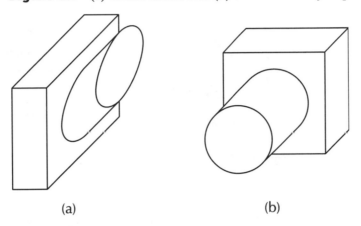

(a)   (b)

Figure 7.6 (a) Inconvenient and (b) convenient object positions

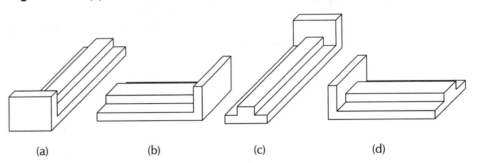

(a)   (b)   (c)   (d)

Figure 7.7 Oblique projection of an object with a long dimension: (a) inconvenient position, (b) view has hidden parts, (c) inconvenient position, (d) best position

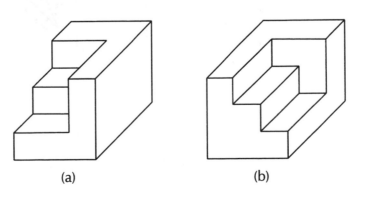

(a)   (b)

Figure 7.8 (a) Inconvenient and (b) convenient object positions

7.3 Oblique projection of lines

As was stated before, front faces and all parallel faces retain their true shape and size in oblique views. Consequently, any lines that lie in these faces or, in other words, are parallel to the projection plane are projected with their true length and inclination. On the other hand, lines in other planes are drawn by locating the projection of their end-points. Figure 7.9 shows how the lines of a geometric object are projected in an oblique view.

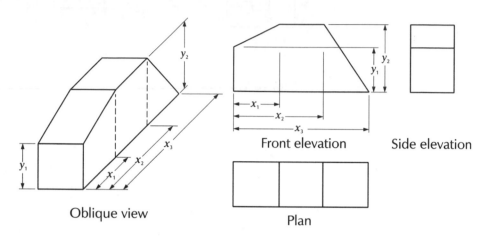

Figure 7.9   Oblique projection of lines

7.4 Oblique projection of angles

Angles that lie in a plane parallel to the projection plane are drawn with their true values as in orthogonal views (see Fig. 7.10a). Angles in other planes are drawn by using the coordinates of points on the two lines that contain the angle as shown in Fig. 7.10b. Alternatively, the tangent value of the angle may be used (see Fig. 7.10c).

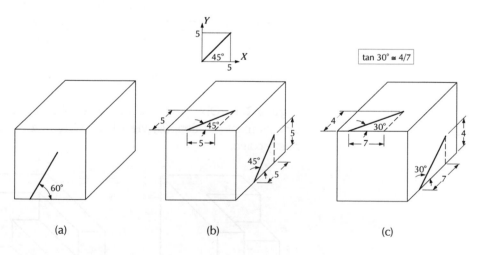

Figure 7.10   Oblique projection of angles. Drawing angles (a) on front face, (b) using point coordinates, (c) using tangent values

7.5 Oblique projection of circles and curves

When a circle lies in a plane parallel to the projection plane it is drawn in its true shape and size. For this reason, it is advisable to position objects in oblique views so that any circular parts are parallel to the projection plane.

Circles that are in the top or side plane, or in any other plane parallel to one of them, are projected as ellipses. For these circles, two methods used to draw their oblique views are presented in the following sections, namely the method of four centres and the method of intersecting lines.

Finally, for a circle whose plane is not parallel to any of the front, side or top face, a general method to draw its oblique view is presented. This method can also be used to project curves in any plane.

Method of four centres (approximate method)

To draw a circle whose plane is parallel to either the top or side face, the steps given below are followed (refer to Fig. 7.11):

1 Draw rhombus ABCD that contains the circle by drawing lines AB and DC parallel to diameter HF, and lines BC and AD parallel to diameter EG.

2 Erect perpendiculars to the four sides of the rhombus at their mid-points and locate their intersection points I, J, K and L.

3 Draw arcs EF, FG, GH and HE whose centres are L, J, K and I respectively to construct an approximate oblique projection of the circle.

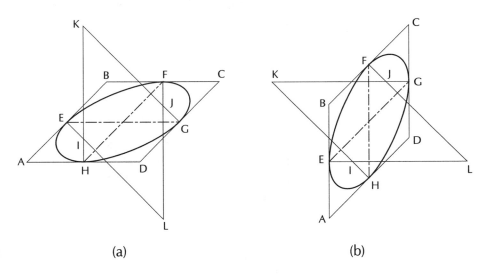

(a)   (b)

Figure 7.11   Oblique projection of circles by the method of four centres. Drawing circles (a) in top plane, (b) in side plane

Method of intersecting lines

A more accurate method to draw an oblique projection of a circle whose plane is parallel to the top or side face is the method of intersecting lines. The method is illustrated in Fig. 7.12 and explained in the following working steps:

1 Draw a similar circle in the front face with the true shape and size.

2 Draw a series of horizontal and vertical lines that intersect at points on the circle perimeter.

3 Project a similar series of lines on the plane of the circle under consideration, and locate the points of intersection.

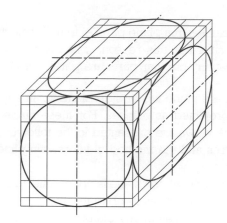

Figure 7.12   Oblique projection of circles by the method of intersecting lines

4 Draw a smooth curve through these points to represent the oblique projection of the circle.

General method of projecting circles and curves

This method may be used to draw an oblique projection of any circle, or curve, regardless of its plane. According to this method, a curve is considered as a series of lines connecting a group of points. Projection of these points can simply be done using their coordinates; then the curve projection is drawn by connecting the projected points by a smooth curve. Figure 7.13 illustrates the use of this method with a typical example.

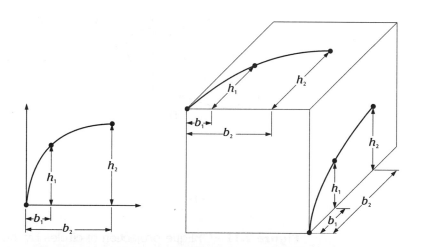

Figure 7.13   Oblique projection of general planar curves

The same principles can be applied to project three-dimensional curves (see Fig. 7.14).

Common errors in projecting circles and curves

The common errors made in constructing oblique views of circular parts are similar to those experienced with isometric views. Reference may be made to the last part of Sec. 6.4 and Fig. 6.8.

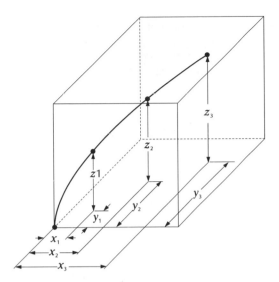

Figure 7.14 Oblique projection of three-dimensional curves

7.6 Oblique projection of three dimensional objects

Having learnt how to draw oblique projections of planar lines, angles and circles, projecting objects becomes a simple step requiring only the consideration of the third dimension. Oblique projection of a three-dimensional object is commonly done through the following steps. Refer to Fig. 7.15 for an example. Also, notice the similarity between isometric and oblique projection of objects by comparing Fig. 6.9 to Fig. 7.15.

1 Position the object while considering the recommendations 1 to 4 presented in Sec. 7.2.

2 Draw the orthogonal views of the object and enclose them in the smallest rectangular box possible. Draw the views to a scale equal to that of the required oblique view.

3 Draw an oblique projection of the box with its front elevation face being the front face of the oblique view.

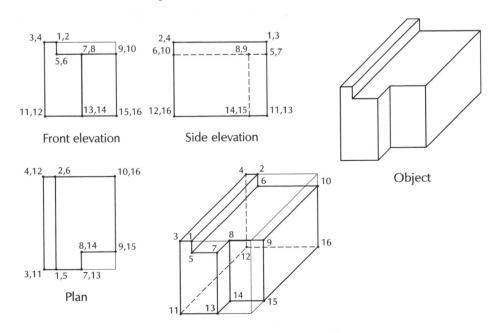

Figure 7.15 Oblique projection of an object

4 Start with the parts that lie at the front face and then proceed away from there. Project points first using their coordinates and then draw the lines lightly.

5 Erase construction lines and any unrequired line extensions and then finish with heavy lining.

6 It is a common practice to omit all hidden lines in pictorial views for clarity. However, this is not a rule and sometimes hidden profiles may be included.

7.7 Disadvantages of oblique projection

The above sections illustrate the efficiency and ease of use of oblique projection which make it one of the most popular projection techniques in civil engineering and architectural applications. However, oblique projection still suffers from a disadvantage which is general to pictorial drawing techniques, that is drawings are incomplete. For example, hidden parts are usually omitted for clarity, and if included they might not, in most cases, be fully dimensioned.

Another disadvantage is the undesirable distortion of the top and side faces (see, for example, Fig. 7.16). The extent of distortion here is worse than that of isometric drawing. This in fact emphasizes the importance of proper object positioning as was discussed in Sec. 7.2. Repositioning the objects of Fig. 7.16 so that the top face of object (a) and the side face of object (b) become the front faces could reduce the problem of face distortion substantially.

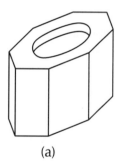   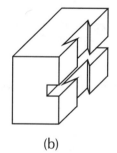

(a)   (b)

Figure 7.16 Distortion of the top and side faces in oblique views

However, unlike isometric views, oblique views project objects in their true sizes without change. For example, spheres that need special treatment in isometric views are drawn in oblique views in a straightforward manner.

Finally, it can be seen that oblique projection is not in fact a true pictorial projection system. If the front face is seen undistorted, as in an orthogonal view, then it should be impossible to see the third dimension of objects. However, this fact is usually overlooked for the sake of convenience and simplicity.

7.8 Dimensioning of oblique drawings

The principles of dimensioning orthogonal views discussed in Sec. 3.2 apply in oblique drawing with the following additions:

1 All dimension and extension lines must lie in the main oblique planes[†], (see Fig. 7.17).

---

†Main oblique planes are any planes that are parallel to the front, top or side face of an ogject

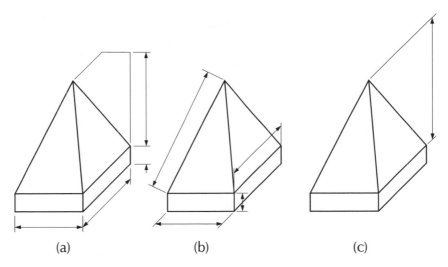

(a)   (b)   (c)

Figure 7.17 Dimension and extension lines in oblique projection: (a) correct, (b) poor practice, (c) incorrect

2 All dimension figures must lie in the oblique planes to which they are related. Alternatively, vertical lettering may be used (see Fig. 7.18). However, consistency must be observed in all cases.

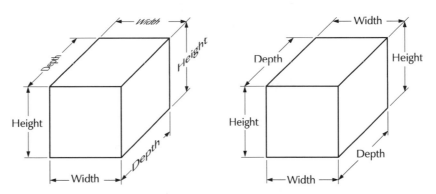

Figure 7.18 Possible systems of dimensioning oblique views

7.9 Exercises   Draw oblique views for the objects presented in Figs 7.19 to 7.34.

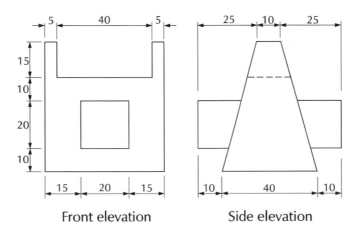

Front elevation   Side elevation

Figure 7.19

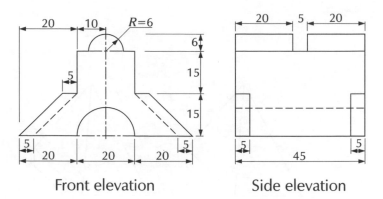

Front elevation   Side elevation

Figure 7.20

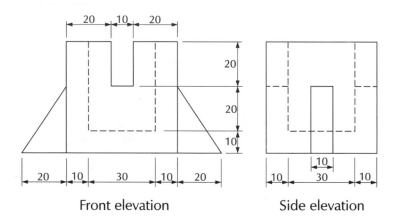

Front elevation   Side elevation

Figure 7.21

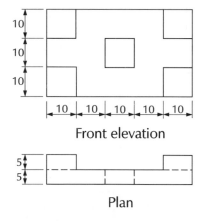

Front elevation

Plan

Figure 7.22

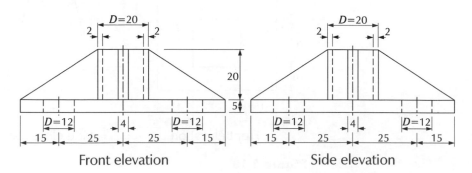

Front elevation   Side elevation

Figure 7.23

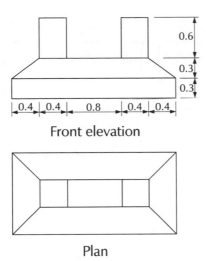

Front elevation

Plan

Figure 7.24

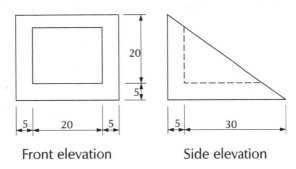

Front elevation   Side elevation

Figure 7.25

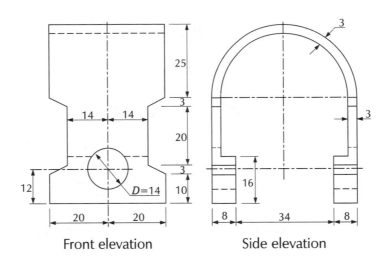

Front elevation   Side elevation

Figure 7.26

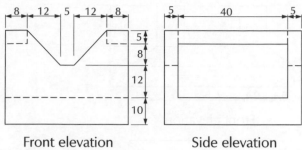

Front elevation   Side elevation

Figure 7.27

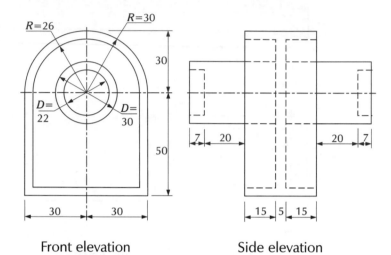

Front elevation   Side elevation

Figure 7.28

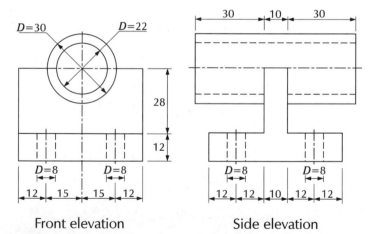

Front elevation   Side elevation

Figure 7.29

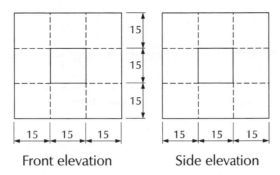

Front elevation   Side elevation

Figure 7.30

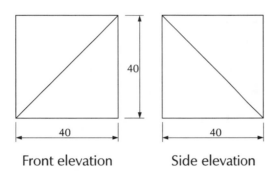

Front elevation   Side elevation

Figure 7.31

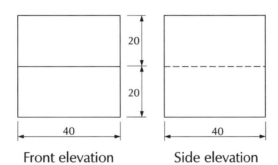

Front elevation   Side elevation

Figure 7.32

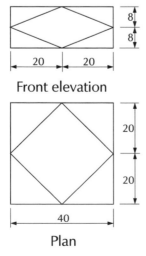

Front elevation

Plan

Figure 7.33

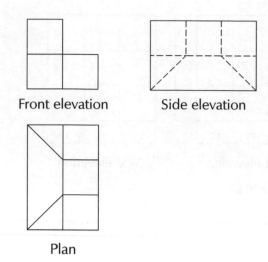

Front elevation   Side elevation

Plan

Figure 7.34

Intersection of Geometric Objects

This chapter introduces the intersection of geometric objects. It is designed to be a reference chapter, referred to only when needed.

8.1 Types of geometric objects

Various types of geometric objects are used in civil engineering projects. They may be divided into the following categories.

Regular geometric objects

These objects have edges of the same length, and include: the tetrahedron (the regular triangular pyramid), the hexahedron (the cube), the octahedron (with eight regular faces), the dodecahedron (with twelve regular faces) and the icosahedron (with twenty regular faces) (see Fig. 8.1).

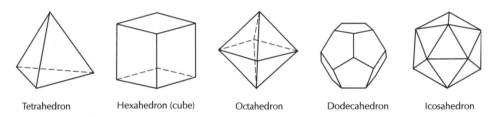

| Tetrahedron | Hexahedron (cube) | Octahedron | Dodecahedron | Icosahedron |

Figure 8.1   Regular geometric objects

Prisms and pyramids

Prisms and pyramids may be right, oblique or truncated, as shown in Fig. 8.2.

Objects with curved surfaces

These may be curved in one direction such as the cone and the cylinder, or curved in two directions such as the sphere, etc. (see Fig. 8.3).

8.2 Intersection of prisms and pyramids with planes

In most cases, the intersection polygon can be determined by locating and connecting the points of intersection of the plane and the edges of the object. Figure 8.4 shows the polygon of intersection of a plane and a pyramid.

Figure 8.5 shows the intersection of a plane and a prismatic object with a constant cross-section.

Figure 8.6 gives two more examples on the intersection of two prisms.

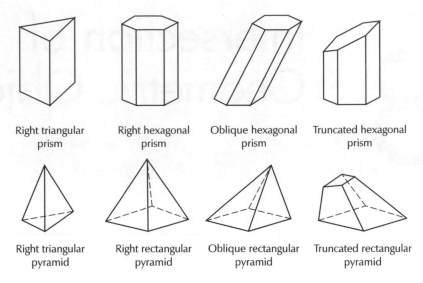

Right triangular prism   Right hexagonal prism   Oblique hexagonal prism   Truncated hexagonal prism

Right triangular pyramid   Right rectangular pyramid   Oblique rectangular pyramid   Truncated rectangular pyramid

Figure 8.2   Prisms and pyramids

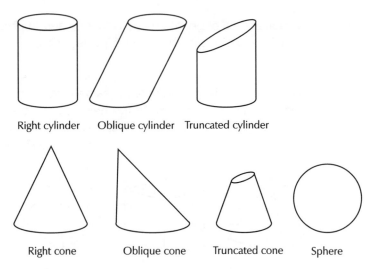

Right cylinder   Oblique cylinder   Truncated cylinder

Right cone   Oblique cone   Truncated cone   Sphere

Figure 8.3   Objects with curved surfaces

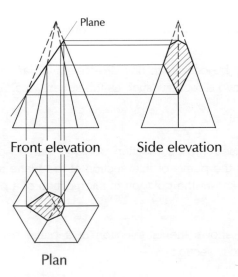

Plane

Front elevation   Side elevation

Plan

Figure 8.4   Intersection of a plane and a pyramid

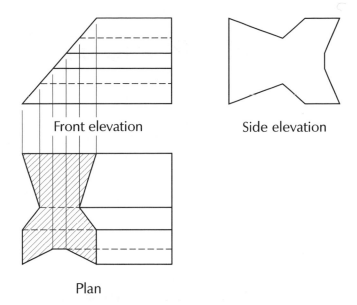

Front elevation   Side elevation

Plan

Figure 8.5   Intersection of a plane and an object with a constant cross-section

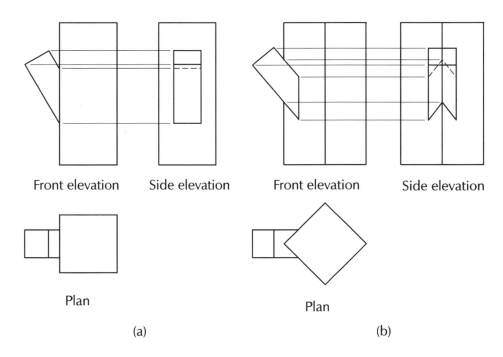

Front elevation   Side elevation   Front elevation   Side elevation

Plan   Plan

(a)   (b)

Figure 8.6   Two examples on the intersection of two prisms

8.3 Intersection of cylindrical objects with planes

A plane intersects with a cylinder in an ellipse whose longest and shortest diameters can be obtained as shown in Fig. 8.7. The ellipse may then be drawn as explained in Sec. 4.8.

Figure 8.8 shows two further examples on the intersection of planes and objects with cylindrical parts.

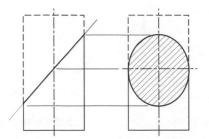

Front elevation   Side elevation

Plan

Figure 8.7 Intersection of a plane and a cylinder

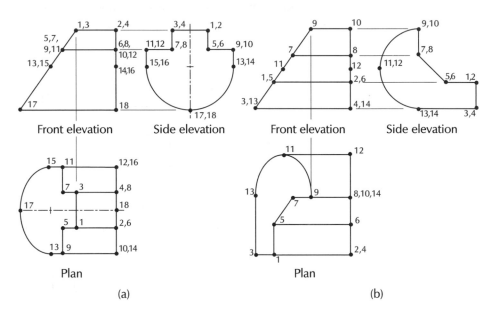

Front elevation   Side elevation   Front elevation   Side elevation

Plan   Plan

(a)   (b)

Figure 8.8 Intersection of planes and objects with cylindrical parts

8.4 Intersection of conical objects with planes

According to the angle between the cone axis and the plane, the intersection may produce an ellipse, a parabola or a hyperbola, as was explained in Chapter 4. Figure 8.9 illustrates an example of this intersection.

8.5 Intersection of geometric objects

The intersection of geometric objects is determined in most cases by using working planes. These planes are chosen with appropriate position and inclination. Their intersection curves with each of the geometric objects are determined, before the points at which the curves intersect are located. With more planes, more points are located and finally the required curve of intersection is drawn. The accuracy and speed of the work depend to a large extent on the choice of these planes. For instance, if the intersection curve of a sphere and a cone with a vertical axis is to be determined, horizontal and vertical planes may be used with the sphere as the intersection curves are

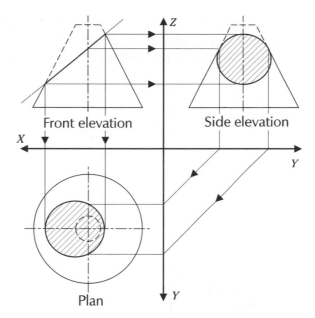

Figure 8.9 Intersection of a plane and a cone

circular in either case, while only horizontal planes may be used with the cone to produce circles; otherwise, difficult-to-handle curves may result.

In general, the working planes recommended for use with the common objects are:

- For a sphere: planes at any inclination;
- For a cone: planes that are perpendicular to the axis to produce circles;
- For a cylinder: planes that are perpendicular to the axis to produce circles, or parallel to it to produce straight lines;
- For a curved surface: planes that are perpendicular to the longitudinal axis.

Among the important examples are:

1. Intersection of two cylinders, the centre lines of which are horizontal and vertical. Horizontal working planes are used in Fig. 8.10. Vertical planes could have also been used.

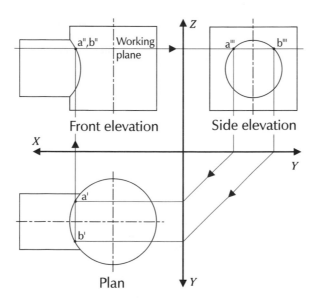

Figure 8.10 Intersection of two cylinders

2   Intersection of two cylinders, one with a vertical axis and the other with a horizontal axis, when the two axes are not in the same plane. Horizontal or vertical working planes could be used in this case. Refer to Fig. 8.11.

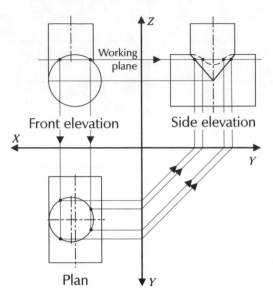

Figure 8.11   Intersection of two cylinders

3   Intersection of two cylinders, one with a vertical axis and the other with an inclined axis, when both axes lay in the same plane. Vertical working planes, parallel to the plane that includes both axes, are used. Refer to Fig. 8.12.

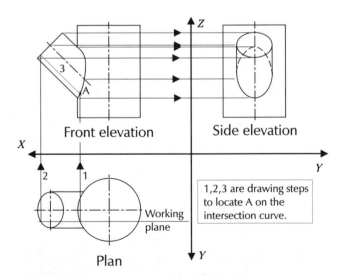

Figure 8.12   Intersection of two cylinders

4   Intersection of a vertical cone and a horizontal cylinder. The working planes must be horizontal, otherwise complex conical sections will result. Refer to Fig. 8.13.

5   Intersection of a sphere and a cone. Working planes that are perpendicular to the cone axis are used in this case. The intersection circles of both the sphere and cone, which correspond to one working plane, intersect at two points on the required intersection curve. Refer to Fig. 8.14.

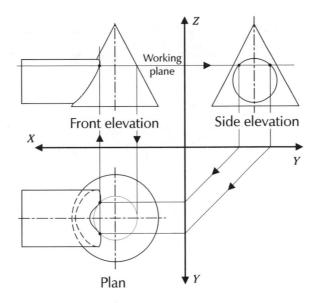

Figure 8.13 Intersection of a vertical cone and a horizontal cylinder

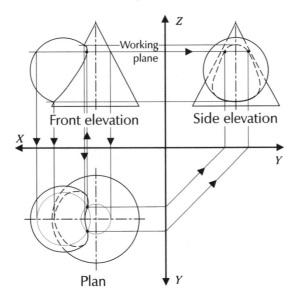

Figure 8.14 Intersection of a vertical cone and a sphere

8.6 Exercises

Draw a third orthogonal view for each of the objects shown in Figs 8.15 to 8.23.

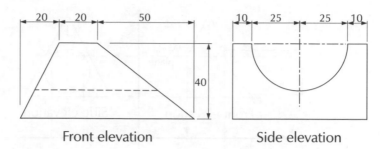

Front elevation   Side elevation

Figure 8.15

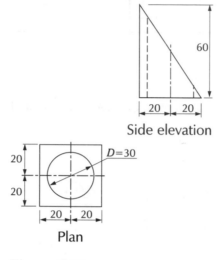

Side elevation

Plan

Figure 8.16

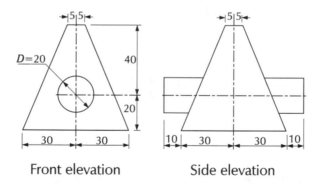

Front elevation   Side elevation

Figure 8.17

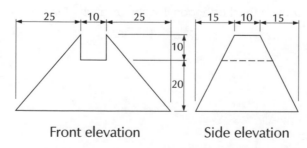

Front elevation   Side elevation

Figure 8.18

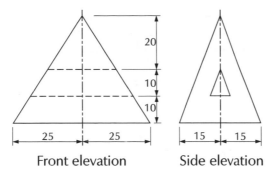

Front elevation   Side elevation

Figure 8.19

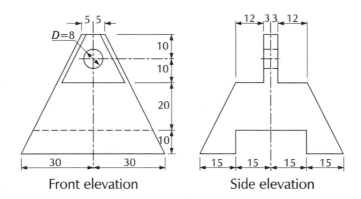

Front elevation   Side elevation

Figure 8.20

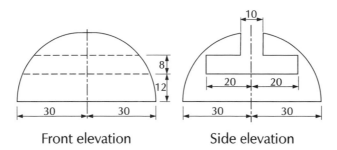

Front elevation   Side elevation

Figure 8.21

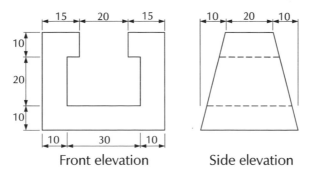

Front elevation   Side elevation

Figure 8.22

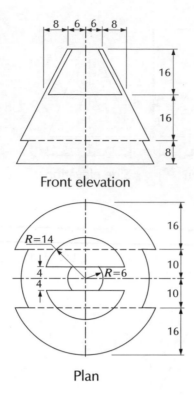

Front elevation

Plan

Figure 8.23

PART 2

Civil Engineering Applications

Before any structure, regardless of its size, cost and importance, can be built, various kinds of drawing are required to convey instructions on the shape, size, materials, colours, etc., of the structure to the constructor. The structure mentioned above may be a house, a factory, a bridge or even a pipeline or a motorway. The drawings required may include layouts, cross and longitudinal sections, details of member connections and general views. As a rule, they must be accurate, complete, precise and neat. They must also be readable without any confusion or doubt. For this purpose, the standard practices and conventions developed by the British Standards Institute, or similar organizations in other countries, for use in the field of draughting must be followed and adopted.

In this part of the book, metallic, reinforced concrete, timber and masonry construction are discussed. The book covers, for every construction material, basic structural elements and basic member connections and then goes on to discuss commonly used structures that are made of these members and connections. Although the book is intended to serve only the draughting aspect of civil engineering applications, hints related to design are given to indicate reasons for choices and preferences, manufacturing and assembly processes, how designs are made safe and when a certain material is best used. All this is explained using simple terms and without unnecessary details. The author believes that this helps draughtsmen grasp the spirit of the structure they are dealing with, and therefore makes draughting more of an interactive process and also less liable to errors.

Throughout this part of the book, examples drawn in orthogonal and pictorial projection views are used to assist learning. Complicated examples are avoided and structures are presented in simple but realistic forms. The same principles illustrated with simple structures also apply to more complicated forms. With the background provided by this book and the experience gained through work, the engineer and the draughtsman will be capable of handling the most complicated forms of structure.

Comprehensive sets of drawings are presented in all chapters of Part 2 as practical examples of typical structures. Not every kind of structure is covered and these examples are introduced only for guidance. The reader is recommended to study these carefully and read the description presented.

CHAPTER 9

Metallic Members and Connections

9.1 Introduction

Metallic structures are mainly made of steel, although aluminium is used in some cases such as in aeroplanes and military bridges. Subsequently, attention is focused in this chapter on steel members and joints. However, it should be emphasized that, from at least the draughting viewpoint, the difference between steel structures and their aluminium equivalents is very limited. The fact that steel is generally stronger, heavier, cheaper and more ductile than aluminium dictates when every material is more adequate, but the detailing of both remains almost the same.

In this chapter, basic steel members and the main methods used to fabricate member connections are presented. Some simple steel joints are also discussed with the help of orthogonal and isometric projection views.

Steel members may be manufactured hot-rolled (extruded through a dye or by calendering) or cold-formed (by folding a steel sheet along certain lines). While hot-rolled members generally have a moderate to high strength, cold-formed members are relatively weaker. Members may also be built up of a number of standard hot-rolled members.

9.2 Hot-rolled steel members

Hot-rolled steel members come in various types and sizes. The main section types commonly in use are as follows (see Fig. 9.1):

(a) Plates, described as PL $a \times b \times t$, where a and b are the length and width of the plate respectively and t is the thickness.

(b) Angles, described as L $a \times b \times t$, where a and b are the widths of the angle legs on the outside and t is the leg thickness. The legs of angle members may be equal or unequal in width.

(c) Universal beams, described as UB $d \times b \times m$, where d is the overall depth of the member, b is the breadth of its flanges and m is the mass in kilograms per metre. These members are mainly used as beam elements.

(d) Universal columns, commonly used as column elements. They are described as UC $d \times b \times m$ as for the universal beam members.

(e) Steel joists. These members are I-shaped, but unlike the UB and UC members, the joist flanges vary in thickness from a maximum at the connection with the web to a minimum at the edges.

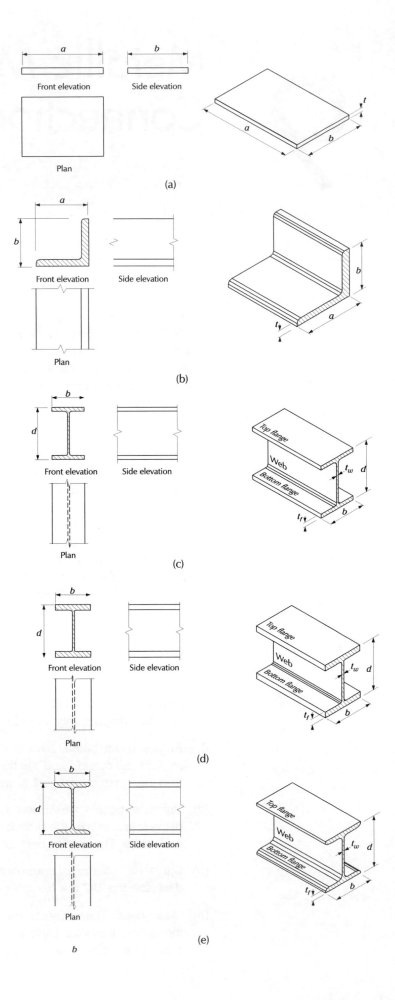

Front elevation   Side elevation

Plan

(a)

Front elevation   Side elevation

Plan

(b)

Front elevation   Side elevation

Plan

(c)

Front elevation   Side elevation

Plan

(d)

Front elevation   Side elevation

Plan

(e)

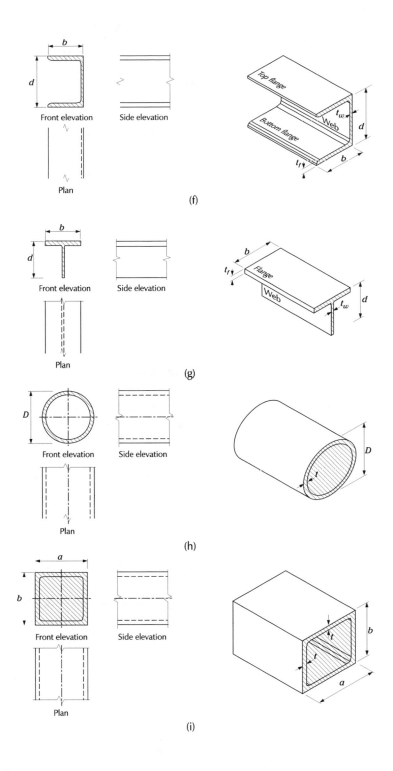

Figure 9.1 Steel standard members: (a) plate PL a×b×t, (b) angle L a×b×t, (c) universal beam UB d×b×m, (d) universal column UC d×b×m, (e) rolled steel joist RSJ d×b×m, (f) channel C d×b, (g) T section T d×b, (h) circular hollow section CHS D×t, (i) rectangular hollow section RHS a×b×t

(f)   Channels, described as C $d \times b$.

(g)   Members of T-section, made by removing a flange and part of the web of a UB or a UC member. T-sections are described as T $d \times b$, where d and b are the overall depth and breadth of the section respectively. The size of the original UB or UC member from which the T-section is made must also be identified.

(h)   Members of circular hollow section, described as CHS $D \times t$, where D is the outside diameter of the section and t is the wall thickness.

(i)   Members of rectangular hollow section, described as RHS $a \times b \times t$, where a and b are the outside dimensions of the section and t is the wall thickness.

The standard sizes of the above members may be found in tables published by steel member manufacturers and construction institutes such as the British Steel Construction Institute.

Built-up members   In spite of the wide variety of standard members available, it is possible to find that the load on a beam or a column is too heavy even for the largest standard member. In such a case, the required member may be built up of a number of standard members so that the required strength and stiffness are achieved. Figure 9.2 shows seven typical examples of built-up beam and column members.

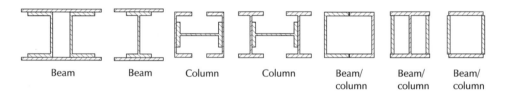

| Beam | Beam | Column | Column | Beam/column | Beam/column | Beam/column |

Figure 9.2   Built-up beam and column members

Built-up members may be shown on construction drawings either as true sections as in Fig. 9.2 for large scale views or diagramatically by heavy lines for small scale views. In the latter method, member parts are drawn separated for clarity (see, for example, Fig. 9.3).

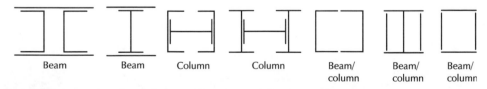

| Beam | Beam | Column | Column | Beam/column | Beam/column | Beam/column |

Figure 9.3   The same members as in Fig. 9.2 but with parts drawn for clarity

9.3 Cold-formed members   Cold-formed members are made by folding steel sheets along certain lines to build members of predetermined cross-sections. Consequently, they may have a wide variety of possible shapes with controllable dimensions (see Fig. 9.4). However, some members can not be produced by cold-forming such as those with I-shaped cross-sections. Also, the sheets used should be within practical thickness limits to enable folding, thus allowing only the production of relatively weak members.

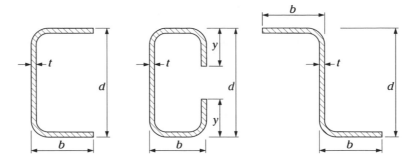

Figure 9.4 Cold-formed members

In construction drawings, long metallic members of uniform cross-section may not be drawn with their true length. They may be broken as shown in Fig. 9.5 so that only a short part of their length is drawn. Figure 9.5 shows how a steel structural member, a tubular member and a cylindrical member are broken.

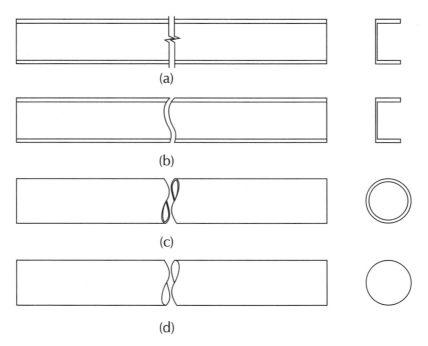

Figure 9.5 Metallic member breaks: (a) and (b) steel structural members, (c) tubular member, (d) cylindrical member

9.4 Metallic joints

Metallic members, made of steel or aluminium, may be joined by means of one of three common methods: riveting, bolting or welding.

Riveted joints

To join, say, two steel plates in a riveted joint, the two plates are drilled at the appropriate positions, and rivets, each with one head and a stem, are inserted through the holes, heated and compressed to form second heads. When cold, the rivets make a firm joint. The holes drilled into the plates must be slightly larger than the rivet cross-section so that rivets can be inserted with ease. The dimensions of a typical rivet are given in Fig. 9.6 in terms of its diameter, D.

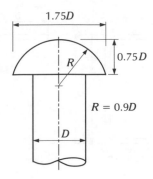

Figure 9.6   Dimensions of a rivet

When more than one rivet is required, rivets are arranged in one or more straight lines with the distance between successive rivets, the pitch, usually constant and equal to three to six times the rivet diameter. The edge distance between the end rivet and the plate edge equals half the chosen pitch and therefore ranges between $1\frac{1}{2}D$ and $3D$.

SPLICE PLATES

One or more splice plates may also be used in forming a riveted joint if the connected members are to be in one plane (see Fig. 9.7). This kind of joint always involves two members carrying the same force and extending in the same direction, and, therefore, usually have the same cross-section.

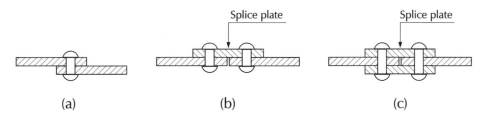

(a)   (b)   (c)

Figure 9.7   Splice plates in riveted joints: (a) no splice plates, (b) one splice plate, (c) two splice plates

GUSSET PLATES

Gusset plates are used to join two or more structural members together when direct connection is not feasible. A typical example is a metallic truss joint where horizontal, vertical and/or inclined members meet. In such a joint, every member is individually connected to the gusset plate which must be strong enough to carry all member forces and transfer them between members. Figure 9.8 shows a typical example of a gusset plate joining a vertical member to a horizontal member. The corners of the gusset plate may be cut out as shown in Fig. 9.8 to reduce the weight. Notice also that the rivets are presented as heavy square crosses. This is the general practice commonly followed in drawing riveted joints (also refer to Fig. 9.10).

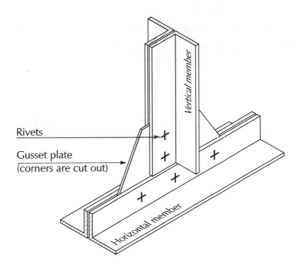

Figure 9.8 A typical example of a gusset plate joint

PACKING PLATES

In some cases, a riveted joint is to be established between two metallic members that have a gap in between (see, for example, Fig. 9.9). A packing plate is needed in such cases to fill in the gap; otherwise the resulting joint will be poor. In drawings, packing plates are hatched with 45° dashed lines spaced uniformly. The packing plate shown in Fig. 9.9 fills the gap between plates 1 and 2. The thickness of the gap, and consequently the thickness of the packing plate, equals the thickness of the angle vertical leg.

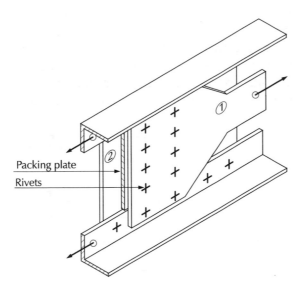

Figure 9.9 A typical example of a joint involving a packing plate

Bolted joints Bolted joints are very similar to the riveted joints discussed above. The bolts used are normally made of a high tensile steel. After being inserted through the holes in the members and screwed into a nut, the bolt is tightened by applying a specific high torque. This produces a high axial force in the bolt and, subsequently, adequate friction between the joined members to prevent relative slip. Bolted joints are more popular than their riveted equivalents, mainly because bolts are relatively easier to fix and can be undone if required.

The detailing of bolted joints is usually identical to that of riveted joints. The only exception is in the representation of bolts and rivets in elevation views. This is illustrated in Fig. 9.10.

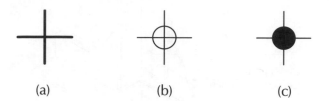

(a)   (b)   (c)

Figure 9.10   Representation of (a) a rivet, (b) a bolt, and (c) an open hole in drawings

Welded joints   Instead of riveting or bolting metallic members together, they may be welded. In this case, there is no need to drill holes in members and connection plates, thus avoiding any reduction in member strength. Welding of joints may be done in a workshop to achieve a high welding quality. Only a few joints are usually left to be welded on site.

The welding technique most commonly used in structural applications is fusion welding, in which a welding rod is melted and combined with the metal parts that are to be fastened together. The melting process can be done using high electric power or torches.

KINDS OF WELDED JOINTS

Welded joints are similar in detailing to bolted and riveted joints, including the use of splice, gusset and packing plates. When two members are to be joined, various kinds of welded joints may be used, the choice of the most adequate depending on the position of one member relative to the other and the required strength. Figure 9.11 shows the main kinds of welded joints, namely the butt joint, the tee joint, the corner joint and the lap joint. The dimensions required for a full description of each joint kind are also given.

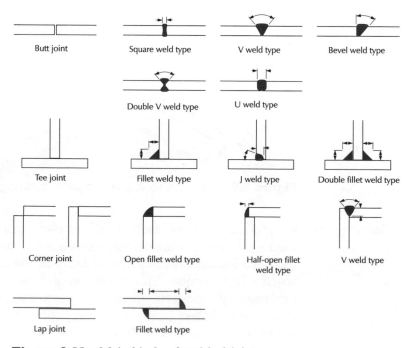

Figure 9.11   Main kinds of welded joints

WELD SYMBOLS

In construction drawings, welding is not shown as in Fig. 9.11, but by means of weld symbols; otherwise drawings would be complicated and difficult to draw and read. The standard symbols used to indicate the welding types are shown in Fig. 9.12.

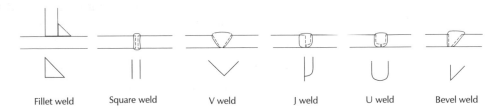

| Fillet weld | Square weld | V weld | J weld | U weld | Bevel weld |

Figure 9.12 Weld symbols

In the following, Fig. 9.11 is reproduced while every welded joint is represented by its symbol. In a typical construction drawing, only the symbol would be used. The standard rules that must be followed in placing weld symbols in order to avoid confusion and wrong interpretation are:

1 A weld symbol is placed above the reference line (the line with an end arrowhead) to describe the weld on the side of the joint away from the arrowhead, e.g. the fillet weld type of tee joint in Fig. 9.13.

2 A weld symbol is placed below the reference line to describe the weld on the same side as the arrowhead, e.g. the V weld, bevel weld and U weld types of butt joint in Fig. 9.13.

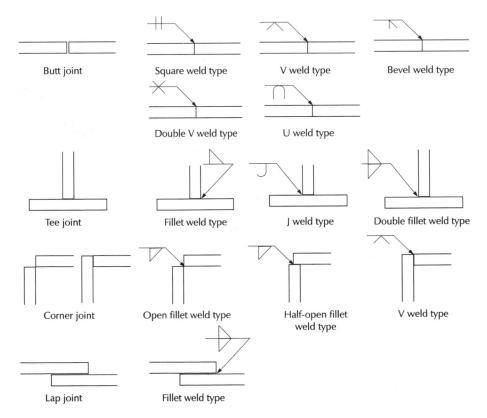

Figure 9.13 The use of symbols to describe welds in metallic joints

3   Weld symbols are placed above and below the reference line to describe the weld on both sides of the joint, e.g. the square weld type of butt joint and the double-fillet weld type of tee joint in Fig. 9.13.

Information regarding the weld size and location can be placed on the reference line. Figure 9.14 shows two typical examples:

1   In Fig. 9.14a, the weld size is 6 mm × 8 mm and the weld is applied for the whole length of the connection. Notice that the weld information is always placed on the same side of the reference line as for the weld symbol.

2   In Fig. 9.14b, the weld size is 6 mm × 6 mm and is applied in strips with 40 mm long each and 70 mm pitch, where the pitch is the centre-to-centre distance between the weld strips.

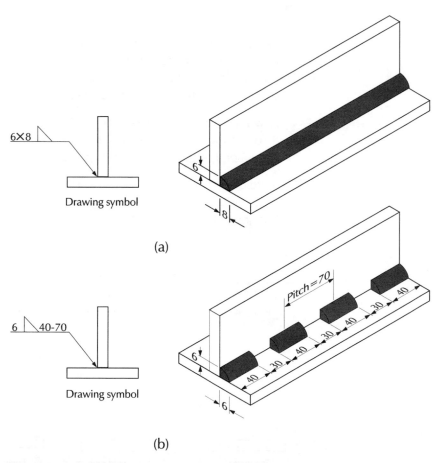

Figure 9.14   Welding information: (a) continuous and (b) non-continuous fillet welds

9.5 Basic connections

In this section, some basic beam and column connections are discussed before a more advanced study on the use of metallic members in construction is presented in Chapter 10. All connections are described with the aid of typical examples presented in orthogonal and isometric views.

On construction drawings, normally two orthogonal views of a connection are sufficient to adequately describe its details. Although isometric drawings are easier to interpret, they tend to be incomplete and are usually used as a supplement to orthogonal views in extreme cases of complexity. On the drawing, the dimensions of the members, connecting angles, connection plates and fasteners are presented clearly and in detail.

Column splice connections

Columns may be spliced to create a hinged connection in a skeletal structure, to ease the assembly process of the structure or when the length of the available column members is not sufficient to cover all the required height. Three of the more common connection configurations are presented in Fig. 9.15. Notice that if the upper column member is smaller in cross-section than the lower column, packing plates may be required to enable the use of splice plates.

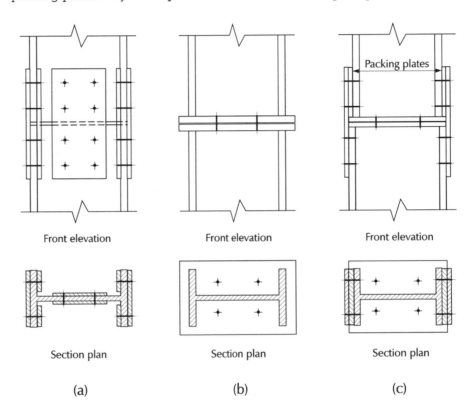

Front elevation　　　　Front elevation　　　　Front elevation

Section plan　　　　Section plan　　　　Section plan

(a)　　　　(b)　　　　(c)

Figure 9.15 Typical examples of column splice connections: (a) fully continuous connection, (b) transfer of forces by bearing, (c) a combination of examples (a) and (b)

Beam splice connections

As for column members, parts of beam members may be connected in splice connections, the most common configurations of which are shown in Fig. 9.16. Although the connections drawn are bolted, they might also be riveted or welded.

Beam/beam connections

The connection between a secondary beam and a main beam may take the form shown in Fig. 9.17a. In this case, the secondary beam spans between the main beams and is supported on their webs by means of connecting angles.

Another form of beam/beam connection is illustrated in Fig. 9.17b, where both beams are continuous through the connection. They are connected in this case by bolts or rivets through their flanges, or alternatively by welding their adjacent flanges together.

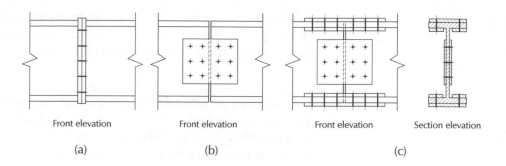

Front elevation   Front elevation   Front elevation   Section elevation

(a)   (b)   (c)

Figure 9.16   Beam splice connections

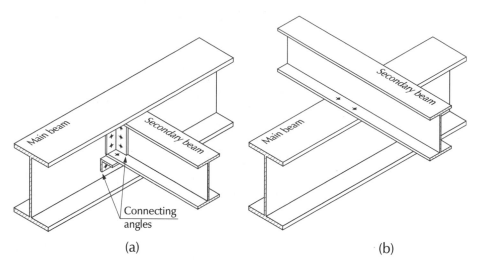

(a)   (b)

Figure 9.17   Typical beam/beam connections

Beam/column
connections

Figure 9.18 shows orthogonal and isometric views of simple and complex beam/ column connections. Beams may be supported on columns by top, bottom and web angles, or by only top angles while sitting on other beams underneath them.

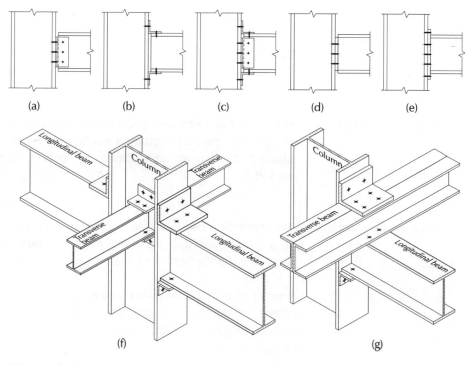

(a)   (b)   (c)   (d)   (e)

(f)   (g)

Figure 9.18   Typical beam/column connections

Column bases
Column bases are used to distribute the heavy column load on to a large area, so that the contact stress under the base is within the bearing strength of the material underneath—be it concrete or soil. Figure 9.19 shows orthogonal and isometric views of typical column bases. The base plate, an essential part of the footing, may be connected to the column directly or by means of side plates and angles. The connection presented in Fig. 9.19c has RHS (rectangular hollow section) bolt boxes covered with plates on the sides of the column. This connection is of particular importance as the side bolt boxes allow the use of long bolts. During connection assembly, the top nuts are turned with a high torque, hence producing a high axial load in the bolts. This helps to create a stiff and rigid connection because even if the bolts creep over time, they will remain under high tension, enough to keep the connection rigid.

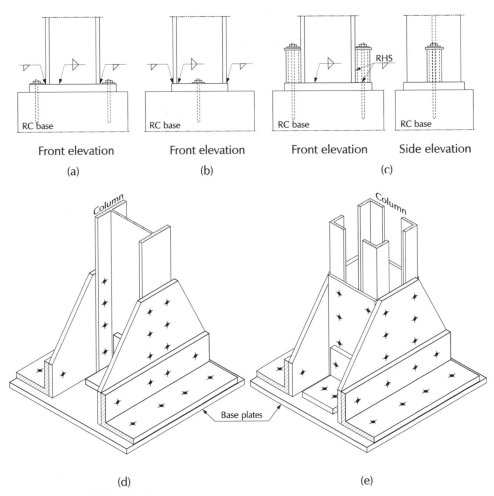

Figure 9.19 Typical column bases

9.6 Exercises   1   For each of the connections shown in Figs 9.20 to 9.30, draw sections AA and BB to a scale 1:10.

2   Draw a pictorial sketch of each of the connections presented in Fig 9.31 and show how the plates are welded together.

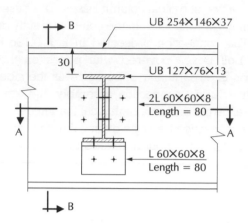

Figure 9.20   Beam/column connection

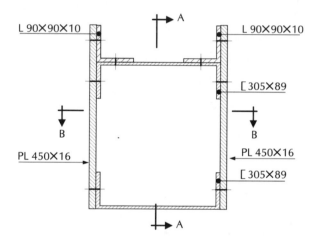

Figure 9.21   Cross-section of a box girder

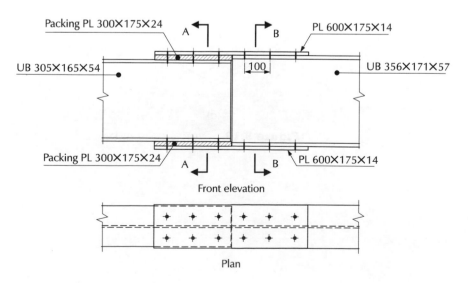

Figure 9.22   Beam splice

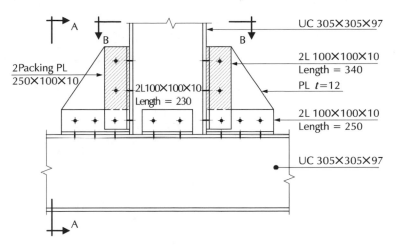

Figure 9.23 Column supported on a beam

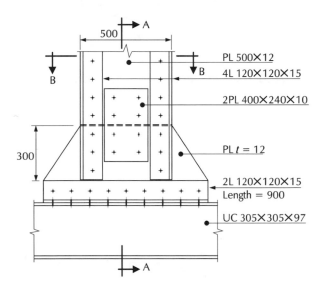

Figure 9.24 Column supported on a beam

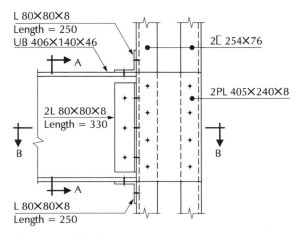

Figure 9.25 Beam/column connection

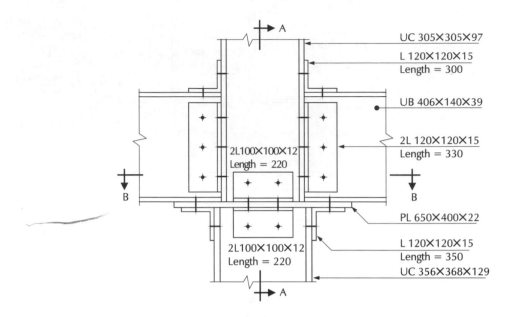

Figure 9.26   Beam/column connection

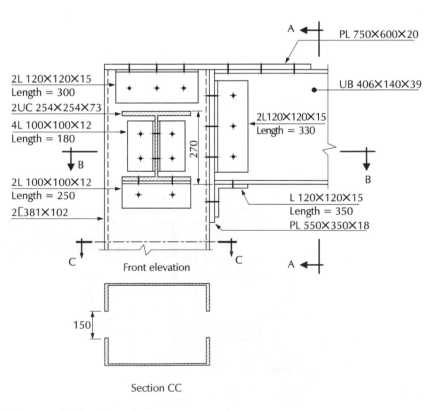

Figure 9.27   Beam/column connection

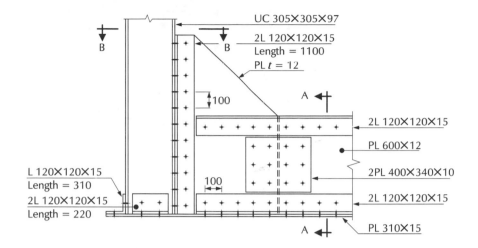

Figure 9.28 Beam/column connection

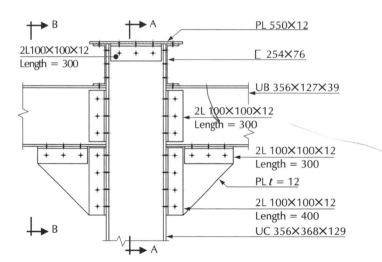

Figure 9.29 Beam/column connection

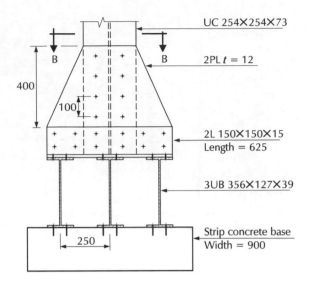

Figure 9.30 Column support

2 Draw a pictorial sketch of each of the connections presented in Fig 9.31 and show how the plates are welded together.

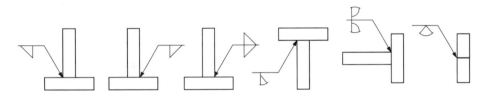

Figure 9.31 Welded connections

CHAPTER 10 Metallic Structures

10.1 Introduction

The eighteenth and nineteenth centuries saw a rapid development in the use of metal in the form of cast and wrought iron for bridges and buildings. It was not until the later half of the nineteenth century that steel was manufactured in sufficient quantity and at a price cheap enough to be a serious competitor to wrought iron.

The production of steel in the twentieth century had allowed bridges and buildings with wide spans and large heights to be built. It is almost impossible to find any building built today that does not use steel in one way or another. Steel is now being used in two- and three-dimensional trusses and frames, lattice bridges, skeletal structures, railways, pipelines and even as reinforcement in reinforced concrete structures. Aluminium is sometimes used to replace steel, for example in military bridges. The same drawing principles for steel construction apply to aluminium structures.

This chapter includes a discussion on the characteristics and detailing rules of steel structures assisted by practical examples of typical structures. However, this chapter must be read in conjunction with Chapter 9 on metallic members and connections. Without a proper understanding of the detailing of these components, overall structures can not be adequately handled.

10.2 Trusses

A truss is a triangulated framework of members in which external loads are resisted by axial forces in the individual members. Trusses may be planar (two dimensional) with all the members in one plane, or three dimensional with members running in three directions. In this section, the discussion is limited to the more common planar trusses.

COMMON FORMS OF PLANAR TRUSSES

Planar trusses are mainly used in buildings to support roofs and floors, commonly spanning large distances, and also in road and railway bridges and footbridges. Planar trusses have various overall forms and internal member arrangements (see, for example, Fig. 10.1).

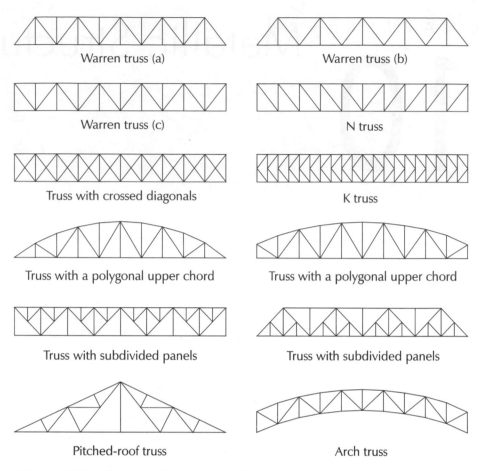

Warren truss (a)

Warren truss (b)

Warren truss (c)

N truss

Truss with crossed diagonals

K truss

Truss with a polygonal upper chord

Truss with a polygonal upper chord

Truss with subdivided panels

Truss with subdivided panels

Pitched-roof truss

Arch truss

Figure 10.1 Common forms of metallic trusses

JOINTS OF TRUSS MEMBERS

Truss members are usually selected from:
- Open sections, e.g. angles and channels;
- Compound sections, e.g. double angles and channels; and
- Closed sections, e.g. hollow rectangular and circular sections.

The pitched roof truss shown in Fig. 10.2 is chosen as an example to illustrate truss member joints. It is assumed that all members are made of angles and all the joints are bolted. Firstly, the details of some chosen bottom and internal joints are presented in Fig. 10.2. It can be seen that while the main members are formed by two angles, secondary members are made of only one angle. Gusset plates are used in all joints. Notice also that main members are continuous through all joints except where it is necessary to terminate a member and start a new one due to member length limitations.

Top truss joints are slightly different. Beams supported by the top chord of the truss and at right angles to it are required to support profiled sheet roofing. These beams, called purlins, have a channel or a Z-section and are fixed to the top chord coincident with the nodal points (see Fig. 10.3).

In some cases, it might be appropriate to replace the commonly used angle members with others of a T-section. In such cases, no gusset plates are required as the diagonal and vertical members are fastened directly to the T-section web (see, for example, Fig. 10.4).

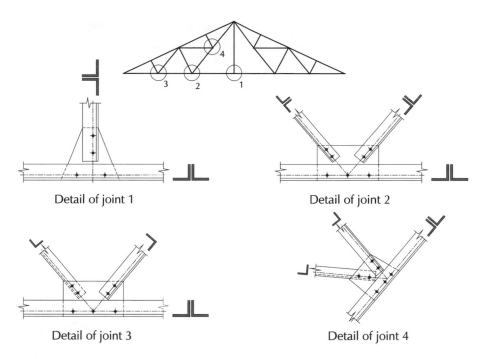

Figure 10.2 Detailing of truss bottom and internal joints

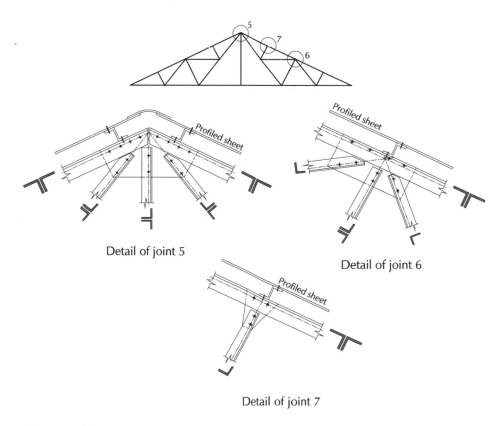

Figure 10.3 Detailing of truss top joints

Members of circular or rectangular hollow sections may also be used as truss members. In such cases, members are commonly welded directly together. Figure 10.5 shows typical connections between members of rectangular and circular hollow sections in a truss.

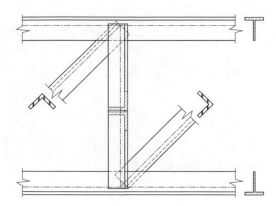

Figure 10.4   Truss with T-section top and bottom chord

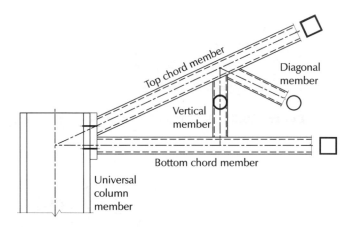

Figure 10.5   Typical connections between members in a truss

TRUSS SUPPORTS

Trusses may be supported on steel columns, concrete columns or walls:

1 Trusses may be supported on steel columns which in turn are supported on ground foundations. The connection between a truss and a steel column could be riveted, bolted or welded, a typical example of which is presented in Fig. 10.6. The connection between steel columns and ground foundations is either fixed or hinged according to the design requirements. This connection

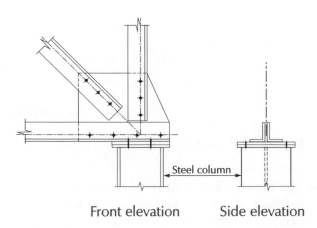

Front elevation   Side elevation

Figure 10.6   Truss supported on a steel column

is similar to that of planar frames studied in Sec. 10.3, typical examples of which are shown in Fig. 10.22.

2   Trusses may also be supported on concrete columns or walls with a hinged or a fixed connection (see Fig 10.7).

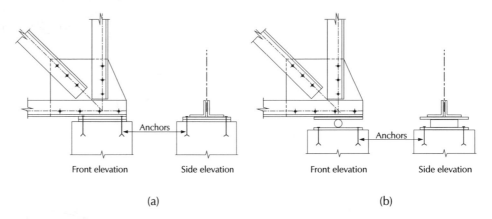

Figure 10.7   Trusses supported on concrete columns: (a) fixed and (b) hinged connections

In some cases, such as a cantilever truss, a fixed end support needs to be simulated. This is normally done by providing a separate support to the ends of both the top chord and bottom chord members (see Fig. 10.8).

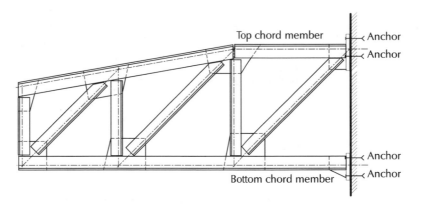

Figure 10.8   Cantilever truss

THE BRACING SYSTEM

Another feature of planar trusses and similar structures is the bracing system. Since these structures can carry loads in only the vertical and longitudinal directions (i.e. in their plane), a separate structural system, the bracing system, must be built to carry forces in the lateral direction, e.g. wind loads. As shown in Fig. 10.9, the bracing system consists of diagonal members joining trusses at the top chord level. For trusses supported on steel columns, other bracing members are used on the sides of the building to carry the lateral forces to the ground foundations. The bracing system is not required between every two trusses and some panels may be left without bracing. In such a case, the purlins help connect the braced bays together.

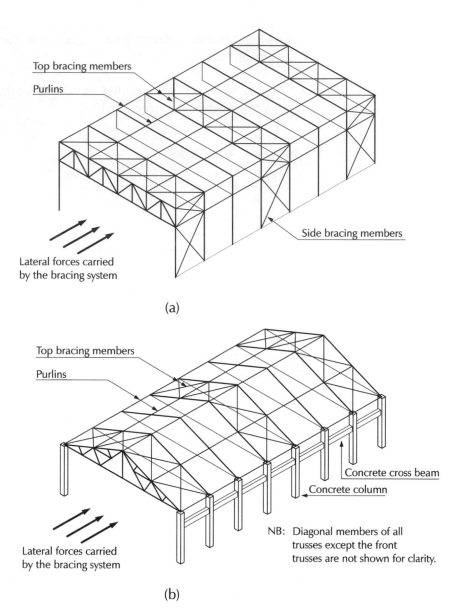

Figure 10.9   The bracing system. Trusses supported on (a) steel and (b) concrete columns

The bracing system is normally constructed of steel bars connected to the main trusses at the top chord level and directly below the purlins. Figure 10.10a shows the details of a typical connection between a bracing bar and a truss top chord. When two bracing bars intersect in the panels between trusses, no special joint is needed, and the bars are simply passed one above the other. In cases when the bracing system is made of rolled sections such as angles, the intersection joints must be carefully designed. Examples of connections of angle bracing members with trusses and at intersection points are presented in Figs 10.10b and c respectively.

In most cases, it is sufficient to detail a truss bracing system on layout orthogonal views of the whole structure. In these views, the truss details are skipped and only the chord centre lines are presented for simplicity. The bracing system is then depicted using thin continuous lines as shown in Fig. 10.11. However, details of connections between bracing members and between them and truss chords must be given on large scale drawings (see, for example, Fig. 10.10).

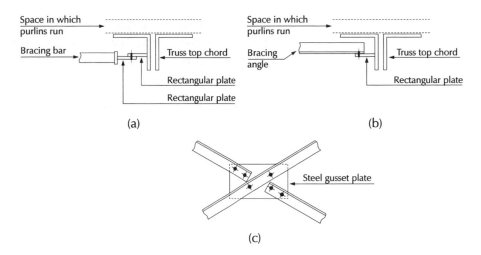

Figure 10.10 Bracing members: (a) end connection of a bracing bar, (b) end connection of a bracing angle, (c) connection between bracing angles

REQUIRED DRAWINGS

In order to fully describe the construction of a truss, the drawings should include:

1   Layout plan and elevation of truss showing all members and supports (see, for example, Fig. 10.11).

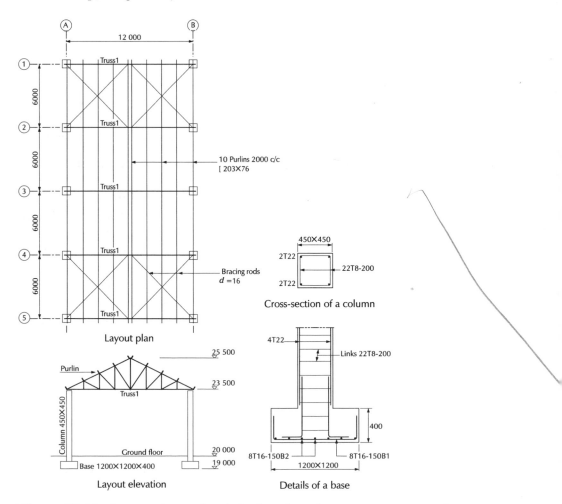

Figure 10.11 Layout views of a pitched-roof steel truss on concrete colums

2   A layout plan giving the location of all bases and columns. The dimensions of all bases and their bottom level must also be presented. These details may be included in the layout views mentioned in part 1 for simple trusses (see, for example, Fig. 10.11).

3   A large-scale elevation of truss showing all members and joints (see, for example, Figs 10.12 and 10.13). Truss supports are also detailed on the same drawing.

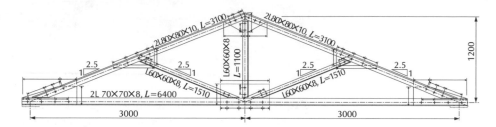

Figure 10.12   The first method of detailing trusses

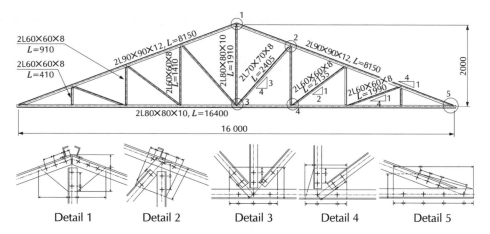

Figure 10.13   The second method of detailing trusses

4   An overall view of the structure to show the bracing system and the purlins. If the elevation and plan views are not adequate, a pictorial view is drawn. The details of the truss might be skipped and only the centre lines of members are drawn. Examples of this are shown in Fig. 10.9.

Two commonly used methods are adopted to detail trusses, either by drawing the truss to a 1:10 scale with all details on or by adopting a 1:25 or a smaller scale for the truss and detailing all joints separately with a 1:5 or 1:10 scale. Every method has its merits and limitations; for example the first method fails with large trusses. Figures 10.12 and 10.13 give typical examples of both methods.

10.3 Skeletal structures

All buildings can be categorized as either wall bearing or skeletal. In wall bearing structures, loads are transmitted through the walls from one floor to another down to the foundations. On the other hand, a skeletal structure has a strong and stiff skeleton of structural members joined together and supported on efficient foundations. All loads including the weight of walls are transmitted only through the skeleton to the ground.

A steel structure's skeleton may have one of two common forms: the unbraced frame system and the braced frame system:

1 An unbraced frame system consists of beams and columns rigidly connected together. This makes the system capable of resisting horizontal forces (e.g. wind forces). Figure 10.14 shows a sketch of a three-dimensional unbraced frame structure with rigid beam/column joints. Unbraced frames may be subdivided into space and planar frames:

(a) Space (or three-dimensional) unbraced frames are unbraced in all directions. They have members running in various directions meeting each other at rigid joints (see Fig. 10.15a).

(b) Planar (or two-dimensional) unbraced frames are unbraced in two directions, but must be braced in the third. A planar frame has members running in one plane and meeting one another at rigid joints. Usually a planar frame structure consists of a series of planar frames supporting secondary beams extending between them (see Fig. 10.15b).

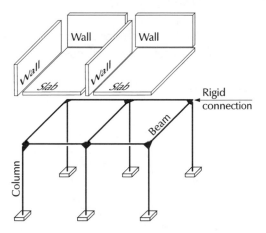

Figure 10.14 Unbraced frame structure

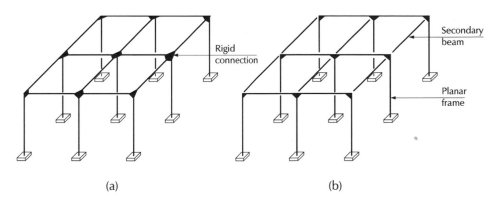

(a)   (b)

Figure 10.15 Unbraced space and planar frame structures: (a) three-dimensional (space) frame, (b) two-dimensional (planar) frames

2 A braced frame system consists of slabs supported by beams which in turn are supported by columns. The columns carry all the structure's weight and loads and transfer them to the ground via the bases. Beam/column connections are designed to be hinges, as shown in Fig. 10.16. This simplifies the design and construction process of the structure but leaves it unable to resist horizontal forces. For this reason, the structure must be

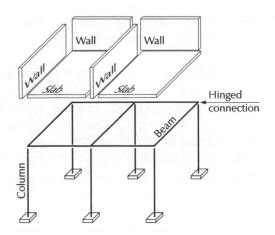

Figure 10.16 Braced frame structure

braced in all directions by using shear walls, diagonal bracing members or strong cores (see Fig. 10.17).

In the following sections, the construction and detailing techniques of unbraced and braced frame structures are discussed.

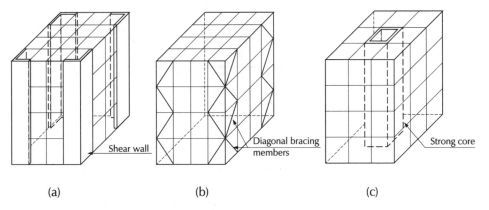

(a)   (b)   (c)

Figure 10.17 Bracing systems of braced frame structures: (a) shear wall, (b) diagonal member, (c) core bracings

Planar unbraced frame structures

The discussion in this section covers only planar unbraced frame structures. The same principles of joint design and detailing apply also to the less used space frame structures.

LAYOUTS OF PLANAR UNBRACED FRAMES

Two-dimensional or planar unbraced frames are of a beam/column construction where most of the beam/column connections are rigid. Planar frames are mainly used to cover moderate spans of about 15 to 30 m. Internally, frame structures are covered with slabs to carry floor loads (in the case of multi-storey frames), while externally they are covered with sheeting and glazing to carry snow and wind loads on the roof and walls.

Planar frames may have any number of storeys and bays (see Fig. 10.18). They are commonly used in industrial sites to cover factories and stores.

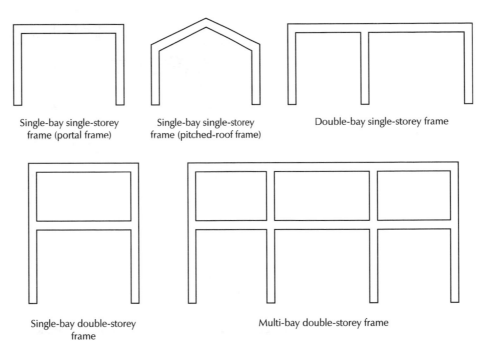

Figure 10.18 Planar frames

BEAM/COLUMN CONNECTIONS

The beams and columns of planar frames are usually made of standard universal beam (UB) and universal column (UC) members respectively. Built-up members may also be used. The beam/column connections are mostly rigid. Figure 10.19 shows seven common ways to produce a rigid frame connection. When the column and/or the beam is continuous in frames with multi storeys and/or bays, beam/column connections may take one of the forms shown in Fig. 10.20.

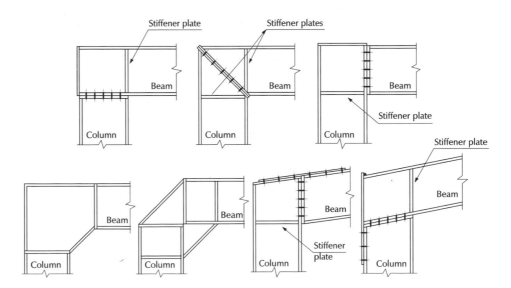

Figure 10.19 Rigid beam/column connections in planar frames

Notice that, in Figs 10.19 and 10.20, additional plates are used while extending between the two flanges of a beam or a column member where a flange of another member is bearing on its side. These plates are called the stiffeners and are used to prevent distortion of the web of one member due to a flow of stress

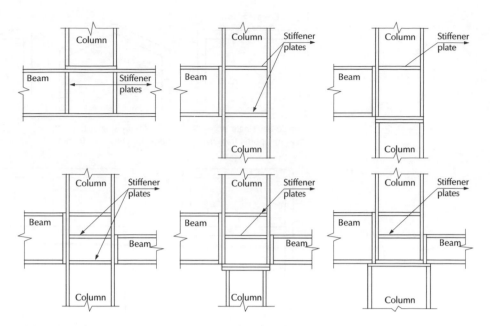

Figure 10.20   Rigid beam/column connections in planar frames

coming from another member leaning on its side. Figure 10.21 illustrates how web plates may be distorted in connections without stiffeners.

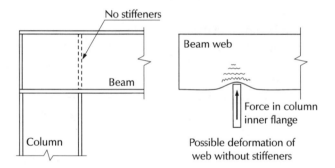

Figure 10.21   Distortion of web plate in an unstiffened connection

FRAME SUPPORTS

The supports of planar frames may be fixed or hinged. Figure 10.22 shows diagrams of typical examples of fixed and hinged supports. Reference is also made to the column/base connections shown in Fig. 9.19.

THE BRACING SYSTEM

It should be noted that a structure that consists of a series of parallel planar frames requires a bracing system similar to that used with planar truss systems (see Fig. 10.23). Steel bars or angles are commonly used as in the case with steel trusses (see Sec. 10.2 and Fig. 10.10).

REQUIRED DRAWINGS

For a complete description of the construction of a planar frame structure, the drawings should include:

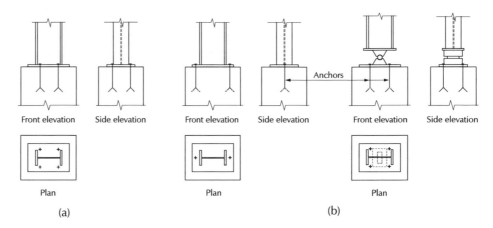

Figure 10.22 Supports of planar frames: (a) fixed and (b) hinged connections

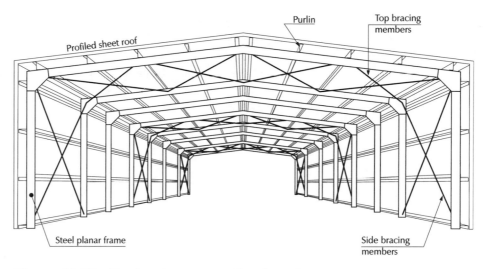

Figure 10.23 The bracing system of a planar frame structure

1 Small scale layout plan and cross-section showing all members, connections and supports. Member sizes and positions and levels of bases and member ends are given on this drawing (see, for example, Fig. 10.24).

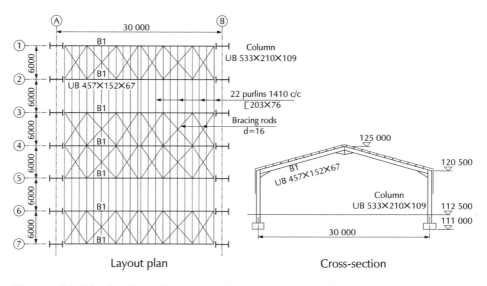

Figure 10.24 Small scale layout plan and cross-section

2 Large scale elevation views of all different members showing length, positions of holes, weldings, stiffeners, end plates, etc. Details of connections to bases are also presented on these drawings (see Fig. 10.25).

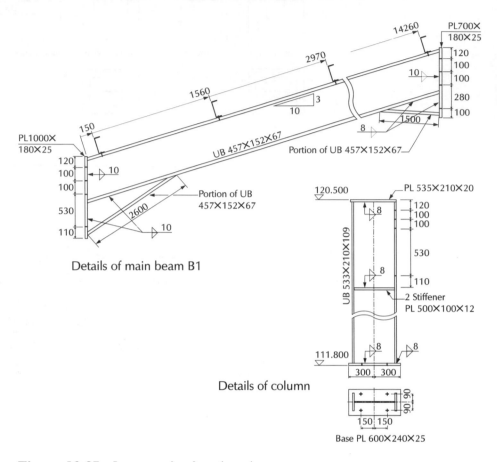

Figure 10.25   Large scale elevation views

3 Overall orthogonal views of the structure showing the purlins and the bracing system. Only the member centre lines need to be drawn on these drawings.

A typical example on the drawings required to detail a planar frame structure is presented in Figs 10.24 and 10.25. Figure 10.24 includes a layout plan and cross-section showing overall dimensions, levels, number of frames and purlins and member sizes. This is followed by a large scale drawing of beam and column members with all the details given, including the details of beam/beam, beam/ column and column/base connections.

Braced frame structures   A steel braced frame structure consists of a three-dimensional skeleton of steel beams and columns joined together in pinned connections with the beams simply supported on the columns. See Fig. 10.17 for sketches of the bracing required for these structures. A comprehensive example on a typical braced frame structure is presented in Figs 10.30 and 10.31.

BEAM/BEAM AND BEAM/COLUMN CONNECTIONS

Figure 10.26 shows examples of main beam/secondary beam and beam/column connections in a braced frame structure (see also Sec. 9.5). Figure 9.18 which

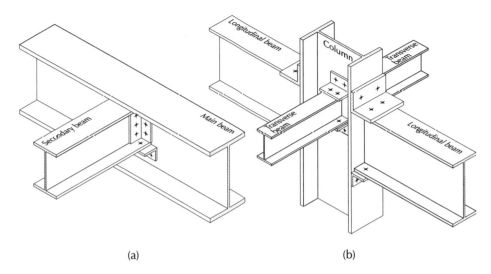

(a)   (b)

Figure 10.26 Member connections in braced frame structures: (a) beam/beam and (b) beam/column connections

presents several configurations of beam/column connections is reproduced here in Fig. 10.27 for easy reference.

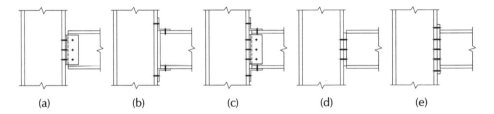

(a)   (b)   (c)   (d)   (e)

Figure 10.27 Further beam/column connections

CONCRETE FLOORS OF SKELETAL STRUCTURES

Floors of metallic braced (and unbraced) structures normally consist of reinforced concrete slabs supported on the beams of the structure's skeleton. To remove the need for a timber formwork to support the concrete slabs during casting, profiled steel decking sheets are laid on the beams, and used to support the slabs, see Fig. 10.28. Unlike the temporary timber formwork, profiled decking sheets are a permanent and integrated part of the structure. Figure 10.29 shows a reinforced concrete slab resting on a profiled decking sheet and the supporting beam.

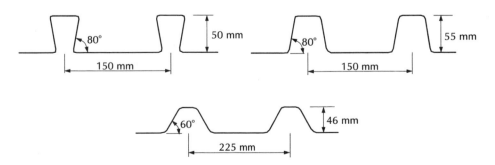

Figure 10.28 Typical forms of profiled sheets

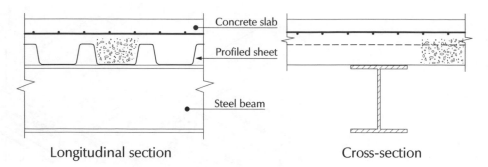

Longitudinal section   Cross-section

Figure 10.29   Concrete slab on a profiled sheet

REQUIRED DRAWINGS

The structural arrangement of steel braced frame structures is usually detailed on plan views where beams may be represented by lines and columns by small scale sections as shown in Fig. 10.30. Notice that each beam is marked with a letter and one or more numbers to indicate the floor (normally A, B, C, ... are used for ground, first, second floor, ... respectively) and the beam serial number. Elevations are also drawn to show the floor heights. The columns are marked according to their corresponding X and Y grid lines. For example, column C2 in Fig. 10.30 is at the intersection of grid lines C and 2.

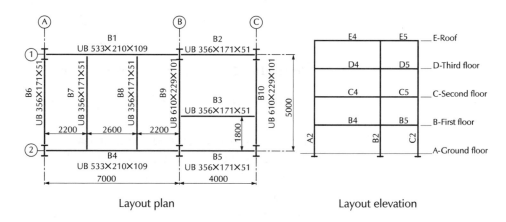

Layout plan   Layout elevation

Figure 10.30   Plan and elevation views of the arrangement of structural members in a multi-storey building

Elevation and section views are also used to detail the columns and beams. Typical examples on the details of columns, main and secondary beams of the braced frame building of Fig. 10.30 are presented in Fig. 10.31.

10.4 Stairs and staircases

Steel may be the most appropriate construction material for staircases in certain cases, as in spiral staircases with small diameter and temporary staircases required during the construction of a building or in industrial plants. Steel staircases are usually cheaper and easier to assemble and build than staircases made of other materials. However, they typically lack a pleasant appearance and may be noisy in use.

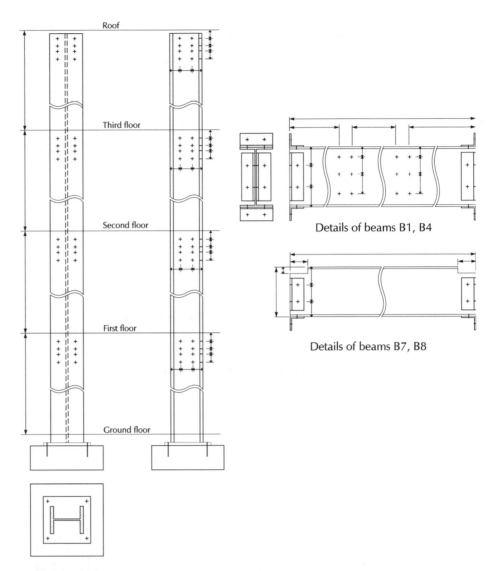

Details of columns A1, A2, C1, C2

Figure 10.31 Details of columns and beams

The simplest form of steel staircase is the rectangular form. Figure 10.32 shows front and side elevation views of a rectangular staircase with one flight. Two channel members are used as main beams with the steps, formed of toes-down channel members, welded to their webs. RHS members are used as vertical rods and handrails.

A more common form of steel staircase is the spiral form, used to reach small places or when the space available for staircase construction is limited. It is normally made for little use and limited to one or two storey heights. A spiral staircase consists of a steel shaft (pipe) fixed at ground or base level, and at the floors it serves. The stairs are of the cantilever form shown in Fig. 10.33. They are rigidly welded to pieces of steel pipes which are then slid on the outside of the shaft. This connection is capable of achieving a stable joint between the shaft and the stairs around it. The stairs have a rough surface to prevent slipping. The handrail is usually a steel strip welded to a set of vertical steel rods, each fitted to the end of a stair. A typical example of a spiral steel staircase is presented in Fig. 10.33.

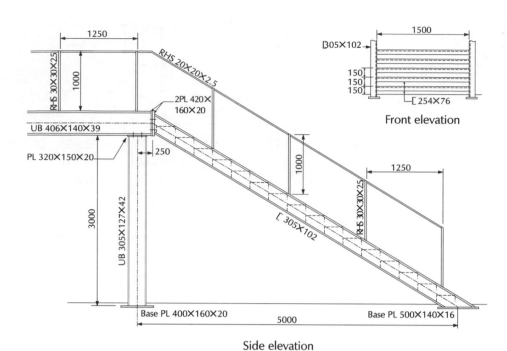

Figure 10.32   Elevation views of a rectangular staircase

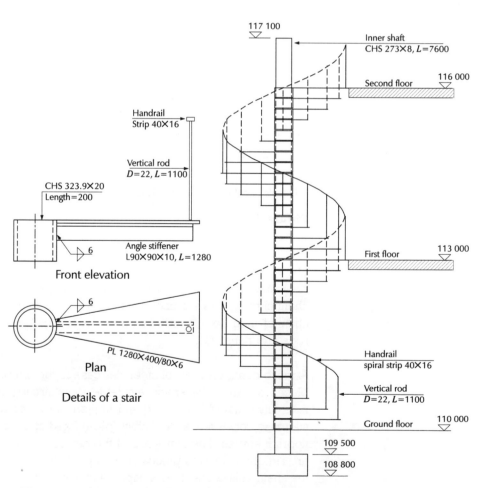

Figure 10.33   Details of a steel spiral staircase

The drawings required to describe the construction of a spiral staircase include:

- An elevation view of the staircase showing the connection between the stairs and the shaft, the base of the shaft and the lateral support provided to the shaft by the different floors;
- Plan and elevation views of a stair showing the connection to the shaft and any stiffeners added on the underside of the stair.

10.5 Exercises

1   For each of the trusses shown in Fig. 10.34, draw layout plan and elevation views showing dimensions, levels, member sizes and positions of trusses. Also show details of all joints on large-scale drawings.

   (a)   Truss (a) has a top chord of double angle L 70 × 70 × 8, a bottom chord of double angle L 60 × 60 × 8 and web members of double angle L 60 × 60 × 6. The structure has seven trusses with 4 m spacing.

   (b)   Truss (b) has a top chord of double angle L 80 × 80 × 8, a bottom chord of double angle L 80 × 80 × 6, verticals of one angle L 80 × 80 × 6 and diagonals of double angle L 80 × 80 × 8. The structure has ten trusses with 5 m spacing.

   (c)   Truss (c) has a top chord and bottom chord of T-section T 100 × 89 (portion of UB 152 × 89 × 16) and web members of double angle L 80 × 80 × 10. No gusset plates are required in this case. The structure has ten trusses with 6 m spacing.

   Assume any required dimensions such as gusset plate sizes.

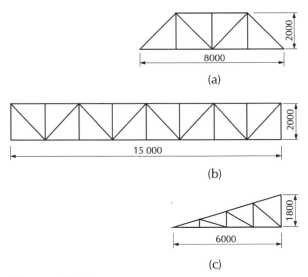

Figure 10.34

2   For the frame sketched in Fig. 10.35 draw layout plan and elevation views showing all dimensions, levels, member sizes and positions. Also draw to a large scale details of column and beam members. The structure has eight identical frames with a 5.5 m spacing. Assume all missing details such as the size and position of purlins and connection plates.

3   The structure whose first floor plan is given in Fig. 10.36 has three storeys, each being 3 m high. All columns are UC 305 × 305 × 137 in the ground floor and UC 254 × 254 × 132 in the two upper floors. Draw a layout elevation of the structure and detail a column, a main beam and a secondary beam to a large scale.

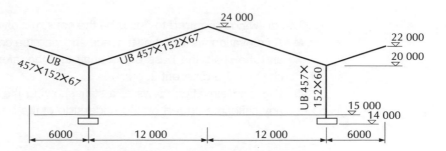

Figure 10.35

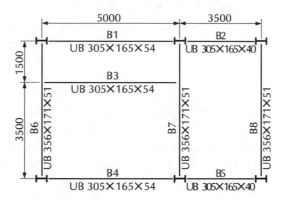

Figure 10.36

CHAPTER 11

Reinforced Concrete Members and Connections

11.1 Introduction

Concrete is arguably the most important construction material, playing a part in most building structures. Its virtue is its versatility and ability to be moulded to take the structural form required. It is also durable, highly fire resistant and cost effective for a wide range of structures. It is in use in all buildings, both single and multi-storey, and may be used in tall buildings with internal cores or wind shear walls.

Concrete is inherently strong under compression but weak under tension. Once loaded in tension, concrete cracks and fails. For this reason, steel bars are added to concrete members where tension stresses are expected to develop. When a concrete member is loaded, the steel bars carry all the tension stresses and reinforce the concrete member against failure. Steel-reinforced concrete members are now so widely used that plain concrete is rarely utilized in a structural component.

Any concrete building can be broken down into the following basic elements: columns, beams, slabs, walls and foundations. These elements are discussed in this chapter along with some typical concrete connections, but, firstly, steel reinforcement, the common factor in all concrete members, is introduced. In the following sections, the methods of detailing concrete members and connections are presented. The reader is also referred to comprehensive sets of drawings of concrete buildings included in Chapter 12.

11.2 Steel reinforcement

A standard range of steel bar sizes is available for use in reinforced concrete. The standard diameters in Europe, Russia and Japan are 8, 10, 12, 16, 20, 25, 32 and 40 mm. Bars of 6 and 50 mm are seldom used. In North and South America, a different range of bar sizes is used, with each size given a certain number. The standard sizes are: #3 (10 mm), #4 (12 mm), #5 (16 mm), #6 (20 mm), #7 (22 mm), #8 (25 mm), #9 (29 mm), #10 (32 mm), #11 (35 mm), #14 (43 mm) and #18 (57 mm).

Steel reinforcement may be bought as individual bars, for use in columns, beams and frames, or in prefabricated mesh form, for use in slabs and walls. The longitudinal and transverse spacings of bars in standard meshes are 200×200, 100×100, 100×200 and $100 \text{ mm} \times 400 \text{ mm}$. The largest diameter of steel bars used in standard meshes is 12 mm, which is adequate for structural purposes.

Two types of bar are commonly in use: high yield steel bars with a yield strength of 460 N/mm2 and mild steel bars with a 250 N/mm2 yield strength. The abbreviated reference letters used on drawings for both types are T and R respectively.

Steel bars must be adequately bonded to the surrounding concrete to form a fully composite structural element. Sometimes steel bars are prepared with straight ends, but more commonly are prepared with either hooks or 90° bends to assist the development of bond.

Cover of steel bars

Steel bars must be embedded in concrete and away from its surface in order to protect them against fire and corrosion. The layer of concrete covering the steel bars is called the cover; its thickness ranges between 20 and 35 mm according to the concrete durability, exposure to moisture and the fire resistance required.

Detailing of reinforcement

Two methods of detailing reinforcement are commonly in use for large and small structures:

1   Reinforcement detailing for large structures. The method commonly used to detail reinforcement in large structures is a tabular method. This method is particularly advantageous when a structure includes a number of elements with a similar profile and reinforcement details, but they differ in dimensions and amount of reinforcement. In this method, a typical element is drawn with reasonable size and each type of reinforcement bars is presented and given a bar mark (a letter and/or a number). A schedule is then included to show, for each individual element and for each bar mark, the corresponding bar shape and dimensions. Bar shapes in this schedule are not described by sketches but by using the bar shape codes identified by BS4466. As an example of the shape codes identified by BS4466, Fig. 11.1 gives a description of three shape codes and the dimensions required to describe each bar shape fully. Figure 11.2 shows the reinforcement details of a solid slab using this method. Notice that symbols B1 and B2 are chosen as bar marks of slab reinforcement bars since, as an additional benefit, they indicate the bottom reinforcement outer layer and inner layer respectively. If required, T1 and T2 may also be used to indicate the top reinforcement outer layer and inner layer.

Shape code	Bar shape	Total length measured along bar centreline	Dimensions required to describe bar shape
20	A	A	A
37	A r B	$A + B - 0.5r - d$	A B
60	A B	$2A + 2B + 20d$	A B

Figure 11.1 Examples of bar shape codes

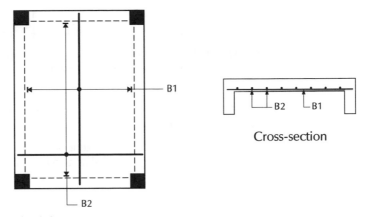

Cross-section

Bar schedule

Member	Bar mark	Type and size	Number of members	Number of bars in each	Total number of bars	Length of each bar (mm)	Shape code	Dimensions required to describe bar shape				
								A	B	C	D	E/r
Slab 12	B1	R10	1	18	18	2400	20	Straight				
Slab 12	B2	R8	1	12	12	3900	20	Straight				

Figure 11.2   Reinforcement details of a solid slab

2   Reinforcement detailing for small structures. When detailing a small structure, or a structure that involves little repetition, it might be more appropriate to put down the reinforcement details on the drawing. An example is presented in Fig. 11.3. In Fig. 11.3, the number, size, type, mark, spacing and position of bars are given. For example, 18R10-1-150B1 means: 18 bars of mild steel and 10 mm diameter laid at 150 mm spacing at the bottom outer reinforcement layer. The bar mark (1 in this example) is a serial number allocated to this bar group on the detail drawing. If, for any reason, another view is to be taken for the same slab, it is acceptable to describe this bar group only by its bar mark. Examples on this are presented in Chapter 12. Using this detailing method, one can differentiate between straight bars and bars with bent ends as in Fig. 11.4a. Also, the arrows are used in different ways to indicate different situations (see Fig. 11.4b).

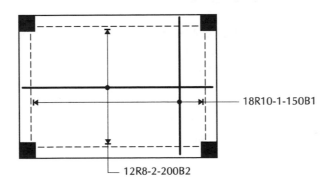

18R10-1-150B1

12R8-2-200B2

Figure 11.3   Reinforcement details of a solid slab

11.3 Columns

Columns are vertical members that primarily carry axial loads. They may have various cross-sectional shapes, such as circular, square, rectangular, L-shaped, box-shaped, etc. As a rule, a longitudinal steel bar must be placed at every corner of the column cross-section. A minimum of six bars is used in circular

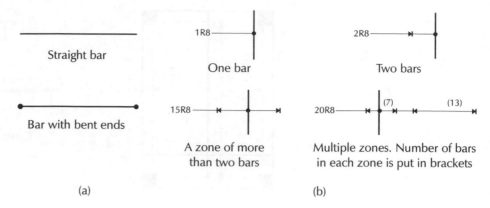

Straight bar

Bar with bent ends

1R8 — One bar

2R8 — Two bars

15R8 — A zone of more than two bars

20R8 — (7)   (13) Multiple zones. Number of bars in each zone is put in brackets

(a)   (b)

Figure 11.4   Representation of steel reinforcement in concrete slabs

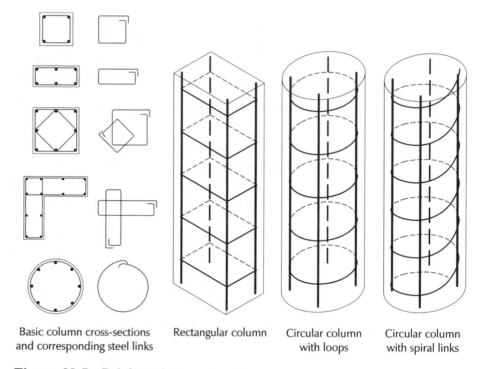

Basic column cross-sections and corresponding steel links

Rectangular column

Circular column with loops

Circular column with spiral links

Figure 11.5   Reinforced concrete columns

columns. Additional bars are required to keep the spacing between bars within 300 mm. The longitudinal bars are contained within steel links (also called stirrups) spaced along the height of the column. Figure 11.5 shows some common column cross-sections and possible arrangements of reinforcement. The figure also shows an alternative way of using links. A steel bar is formed as a spiral (or a helix) and assembled so that it contains the column longitudinal bars. In this case, the column is said to be spirally reinforced.

A typical example on the detailing of concrete columns in buildings is given in Sec. 12.2 and Fig. 12.7.

11.4 Beams

Beams are horizontal members mainly carrying vertical loads. Beams are usually supported on columns and/or main beams, and carry floor slabs and partition walls. When the floor is cast *in situ*, the top part of the beam becomes embedded in the slab, while in precast construction, beams and slabs are

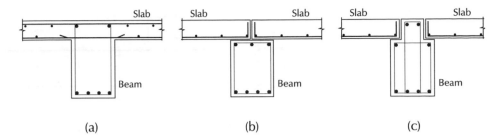

Figure 11.6 Beam/slab connections in concrete floors: (a) concrete cast *in situ*, (b) and (c) precast concrete units

separate. In either case, the part of beam projecting below the bottom surface (soffit) of the slab is usually rectangular (see Fig. 11.6).

Beams have reinforcing bars running longitudinally near their top and bottom surfaces, and normally, if the depth exceeds 750 mm, additional longitudinal bars running along the beam sides are also needed (see Fig. 11.7b). The longitudinal bars are always contained within steel links (also called stirrups) spaced along the beam length. Figure 11.7 shows some typical cross-sections of reinforced concrete beams. The sections shown in Fig. 11.7a, b and c are of beams cast monolithically on site with the slab. Figure 11.7d shows another beam cast with the slab, but because the floor bottom surface ought to be flat in this case, the beam is projected above the slab. Finally, Fig. 11.7e shows a precast I-shaped reinforced concrete beam. Notice that links are used to cover all faces of the beam, and a longitudinal bar is placed at every corner of the beam section and at every point at which the links meet. The beam shown in Fig. 11.7e is normally used in bridges and heavily loaded floors with large spans.

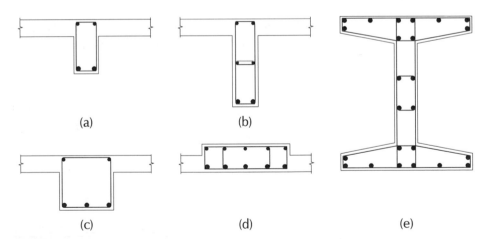

Figure 11.7 Cross-sections of some typical concrete beams

A longitudinal section of a typical beam reinforced with straight bars and uniformly spaced links is shown in Fig. 11.8a. In some cases, design may dictate a non-uniform distribution of links, so that they may be spaced closer together near the supports and wider apart in the middle section (see Fig. 11.8b).

Beam links may take one of several shapes according to the state of stress of the beam and its relative dimensions. The most common link configurations are shown in Fig. 11.9.

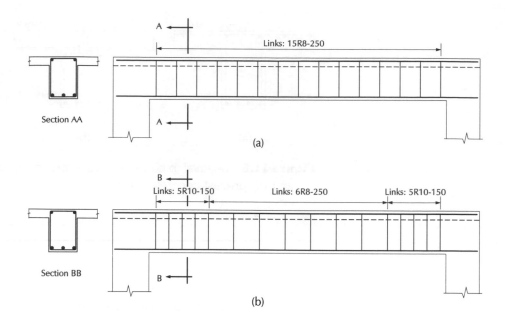

Figure 11.8 Longitudinal distribution of beam links: (a) uniform and (b) non-uniform

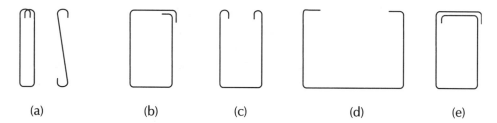

Figure 11.9 Common beam links: (a) links for single bar beams, e.g. ribs of hollow slabs, (b) closed link, (c) open link, (d) open link for wide beams, (e) torsion link

A typical example on the detailing of continuous reinforced concrete beams is given in Fig. 11.10. Notice how bar layers are bent to avoid congestion at the points where bars meet.

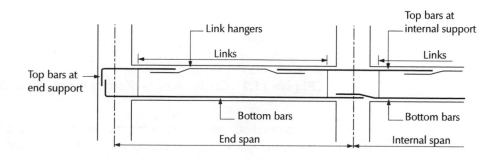

Figure 11.10 Reinforcement details of a continuous beam

Further examples on the detailing of concrete beams are presented in Sec. 12.2 and Fig. 12.10.

11.5 Slabs

Introduction Slabs are horizontal plate elements forming floors and roofs in buildings and normally carry lateral loads. The supporting conditions of slabs vary from one slab to another:

1 A cantilever slab is fixed along one edge (Fig. 11.11a).

2 Slabs may span one way between two beams or two walls (Fig. 11.11b).

3 Slabs may be supported along three sides (Fig. 11.11c).

4 Two-way slabs are supported along all four sides (Fig. 11.11d).

5 Flat slabs are supported directly on columns without beams (see Fig 11.11e).

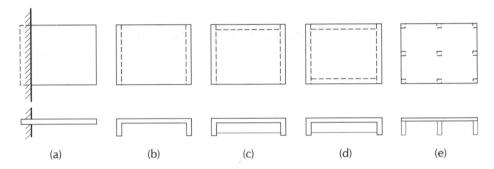

(a)   (b)   (c)   (d)   (e)

Figure 11.11 Types of slab: (a) cantilever slab, (b) slab supported on two beams, (c) slab supported on three beams, (d) slab supported on four beams, (e) flat slab

The reinforcement of slabs is in the form of steel bars running in two perpendicular directions and placed near the slab bottom surface. Only at the internal supports do slabs have reinforcement near their top surface. The only exception is for cantilever slabs where main reinforcement is placed near the top surface. Figure 11.12 shows three cross-sections of (a) a slab simply supported on two edge beams, (b) a slab continuous over three beams and (c) a slab with a cantilever. No top reinforcement is needed in the first case, while top bars are used at the internal support of the second slab. In cantilever slabs such as that shown in Fig. 11.12c, top and bottom bars are used. Bars running in a perpendicular direction to the projection plane appear in the section view as dots. Notice that an adequate concrete cover must be provided between the steel bars and the slab outer surfaces.

Notice also that if another sectional view of the slab shown in Fig. 11.12a is to be taken with respect to a perpendicular projection plane to that used in the above figure, section 2-2 of Fig. 11.13 will result in which the line bars are above the dot bars. It is important to realize that the line bars of section 1-1 are the dot bars of section 2-2, since the two sections are taken with a right angle in between. This in fact is a common error in the drawing of concrete slabs with draughtsmen mixing between the reinforcement layers, sometimes drawing the line bars below the dot bars in two perpendicular sectional views.

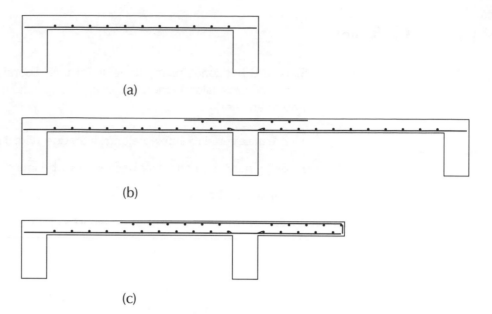

(a)

(b)

(c)

Figure 11.12 Reinforcement details of concrete slabs. Sections through (a) a single-span slab, (b) a double-span slab and (c) a single-span slab with a cantilever

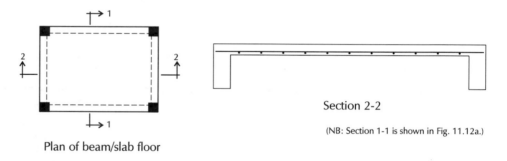

Plan of beam/slab floor

Section 2-2

(NB: Section 1-1 is shown in Fig. 11.12a.)

Figure 11.13 Section through a concrete slab

Slab types according to the cross-section

In general, slabs may be divided according to their cross-sections into solid slabs and ribbed slabs. Both types may be supported on beams or directly on columns.

SOLID SLABS

Solid slabs such as those shown in Figs 11.12 and 11.13 have a constant thickness and uniformly distributed reinforcement bars. This is the most common type of slab, usually used for short to medium spans of up to about 8 m if supported on beams and 6 m if supported on columns (flat slabs). In most cases, a plan view is adequate to show the slab thickness and its reinforcement. The slab shown in Fig. 11.14 has a 200 mm thickness, 24 bars of 16 mm diameter running in the short direction with a 200 mm spacing, and 14 bars of 12 mm diameter and 240 mm spacing running in the long direction. B1 and B2 symbols given to the two groups of bars indicate that the 16 mm bars form the outer layer of reinforcement (nearer to the slab soffit) while the 12 mm bars are laid on them. The arrows pointing at the vertical and horizontal directions beside the slab thickness mean that both directions of the slab, and hence both groups of bars, are effective in resisting the loads applied on the slab.

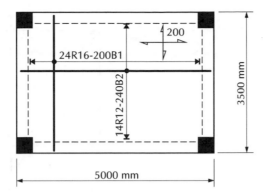

Figure 11.14   Reinforcement of a concrete slab

It is important to note that it is a normal practice to place the bars running in the short direction nearest to the slab outer surface as in the case of the slab in Fig. 11.14.

RIBBED SLABS

To reduce the weight of a solid concrete slab, hollows can be formed into it. This can be done without a significant reduction in slab strength. In this case, the steel bars, previously distributed along the slab side length, must now be grouped in the ribs.

Ribbed slabs may have ribs running in one direction, normally the short direction, and thus called one-way ribbed slabs, or they may be two-way ribbed slabs with ribs running in two directions at a right angle (see Fig. 11.15). Notice that the ribs have links containing their top and bottom reinforcements.

Slab types according to the supports

SLABS IN BEAM/COLUMN CONSTRUCTION

In the examples on solid and ribbed slabs discussed above, the slabs are supported on beams which in turn are supported on columns. This is by far the most common construction system as it is stiff, reliable and safe. A typical example of this system is presented in Sec. 12.2 and Figs 12.6 to 12.10.

FLAT SLAB SYSTEM

Flat slabs, comprising slabs supported directly on columns, are also a popular system of construction. As no internal beams are used, all the floor height to the soffit of the ceiling slab can be efficiently utilized. This makes the system favourable in structures such as car parks and open-space office buildings which may contain multi-directional services in a suspended ceiling void. The slab itself may be solid or ribbed, and the column may be directly connected to the slab or prepared with a column head and/or slab drop (see Fig. 11.16).

Unlike solid slabs on beams, a solid flat slab does not have a uniform distribution of reinforcement. Commonly, solid flat slabs have a high concentration of reinforcement close to the columns and less in the middle of panels. For this

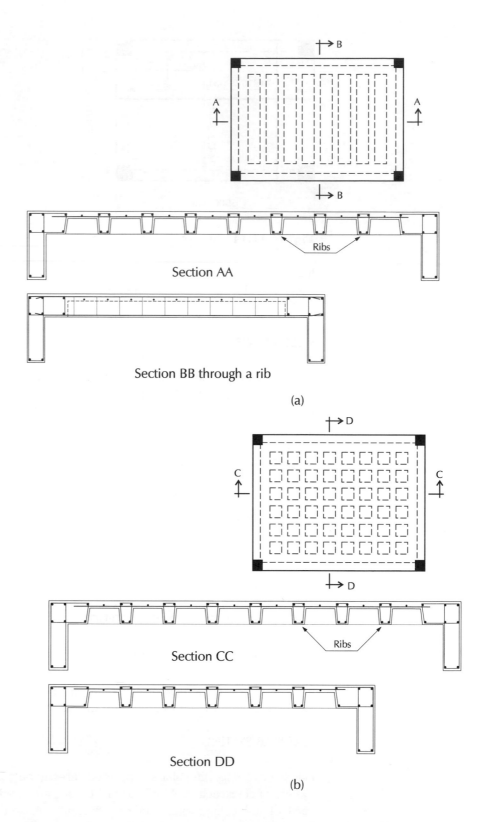

Figure 11.15 Ribbed slabs: (a) one-way ribbed slab, (b) two-way ribbed slab

reason, the slab is divided into column strips and middle strips as shown in Fig. 11.17. This figure presents a typical example of the reinforcement detailing of flat slabs. Symbols B1, B2, T1 and T2 used in Fig. 11.17 indicate the following: B1 ≡ outer layer of bottom bars, B2 ≡ inner layer of bottom bars, T1 ≡ outer layer of top bars and T2 ≡ inner layer of top bars.

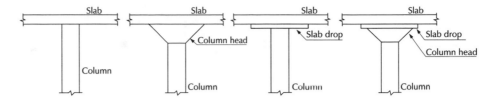

Figure 11.16 Column/slab connections in flat slab structures

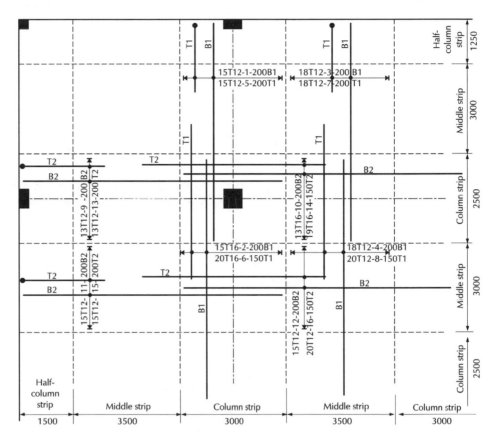

Figure 11.17 Reinforcement of a flat slab

On the other hand, ribbed flat slabs may take one of the forms shown in Fig. 11.18. In Fig. 11.18a, the critical parts of the slab that extend between the columns are kept solid for strengthening purposes and to enable additional reinforcement to be accommodated. The system shown in Fig. 11.18b is not as adequate but could nevertheless be used if the narrow ribs joining the columns are sufficiently strong and stiff. Detailing of ribbed flat slabs is similar to that of solid flat slabs shown in Fig. 11.17.

In flat slab systems, only beams along the edges of the slab are permitted. They help to strengthen the most critical connections between the slab and the edge columns, and carry the external wall. These beams are called marginal (or spandrel) beams.

11.6 Walls

Walls are generally vertical members that carry vertical and lateral loads and have continuous supports. As specified in BS8110, the length of a wall exceeds four times its thickness; otherwise it is classified as a column. Most buildings

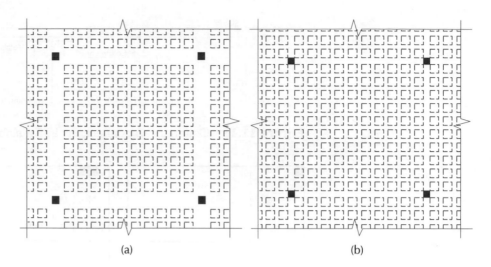

Figure 11.18 Ribbed flat slabs. Columns joined by (a) wide and (b) narrow ribs

contain concrete walls, the functions of which are to carry loads, enclose and divide space or exclude weather and retain heat.

Loads on walls
Reinforced concrete walls may be used in a variety of structures extending from water tanks and earth retaining structures to tall buildings. Loads are generally applied to walls in the following ways:

1   Vertical loads from floor slabs and beams that are supported by the wall (see Fig. 11.19).

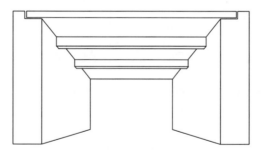

Figure 11.19 Beam/slab floor supported on walls

2   Lateral loads from wind pressure (in external walls of buildings and perimeter walls), water pressure (in water tanks) or earth pressure (in earth-retaining structures) (see Fig. 11.20).

3   Horizontal in-plane loads from wind on, for example, shear walls in a tall building. As an example, the tall building shown in Fig. 11.21 has a narrow width. Reinforced concrete walls are used in this case to stabilize the structure against the horizontally applied wind loads.

Wall details
Concrete walls may retain a constant thickness or the thickness may vary from a minimum at the top to a maximum at the base. Walls are normally reinforced with two layers of steel bars, each having bars running in both the horizontal and vertical direction. Figure 11.22 shows some typical cross-sections of concrete walls and the corresponding bar arrangement.

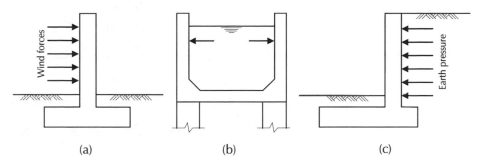

Figure 11.20 Examples of reinforced concrete walls: (a) perimeter wall, (b) water tank, (c) earth-retaining wall

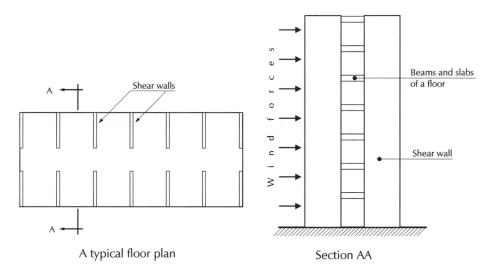

A typical floor plan   Section AA

Figure 11.21 Shear walls in a tall building

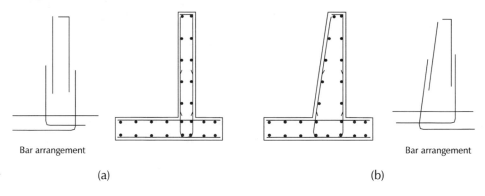

Bar arrangement   Bar arrangement

(a)   (b)

Figure 11.22 Sections through two reinforced concrete walls: (a) wall with a constant thickness, (b) wall with a variable thickness

The arrangement of horizontal bars at the ends of walls and at wall junctions is shown in Fig. 11.23.

Detailing of wall reinforcement

Two typical examples of the detailing of concrete walls are presented in Fig. 11.24. The walls are (a) a wall with a vertical main direction (e.g. a cantilever wall supported by a horizontal base such as those shown in Fig. 11.22) and (b) a wall with a horizontal main direction (e.g. a wall carrying a horizontal load—water or earth pressure—and supported on vertical members such as other walls or piers). In Fig. 11.24, N1, N2, F1 and F2 are used to indicate the following:

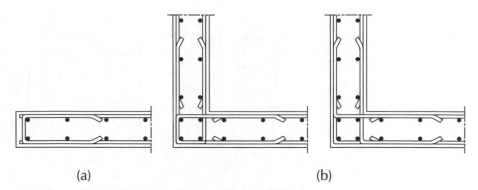

Figure 11.23 Details of walls at ends and junctions: (a) reinforcement details at end of walls (b) reinforcement details at junctions between walls

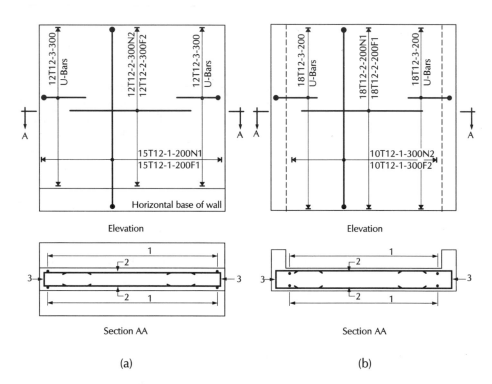

Figure 11.24 Reinforcement details of two concrete walls: (a) wall with a vertical main direction, (b) wall with a horizontal main direction

N1$\equiv$bars at the near face outer layer, N2$\equiv$bars at the near face inner layer, F1$\equiv$bars at the far face outer layer and F2$\equiv$bars at the far face inner layer. Notice that in plan views, only the bar marks are used as the wall reinforcement is fully described in the elevation views.

11.7 Bases

Bases are pads supported directly on the ground while spreading the loads from columns and walls so that they can be supported by the ground without excessive settlement. Alternatively, bases may be supported on piles. The bases discussed in this section can also be used as footings of steel, timber or masonry columns with no change, except in the connection between the column and the footing. For details of this connection, refer to Secs 10.3, 13.4 and 14.3.

Concrete bases may be categorized as follows:

1 Isolated bases for individual columns (see Fig. 11.25). In all cases, the base is reinforced with bottom steel bars running in the longitudinal and transverse direction. Notice that the steel bars running in the short direction are placed nearest to the base bottom surface. Also, a plain concrete base may be cast under the reinforced concrete base if necessitated by the soil condition.

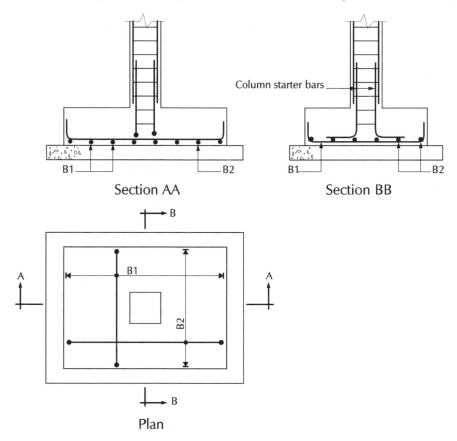

Figure 11.25 An isolated column base

Column starter bars are also inserted in the base before casting. When the column is constructed at a later stage, its longitudinal bars are spliced to these bars to guarantee a good column/base connection. The column/base connection depicted in Fig. 11.25 is fixed. A pinned connection may also be constructed as shown in Fig. 11.26. This connection is essential for planar frames where a hinged support is a requirement in some cases.

Individual bases are detailed using a plan and a section elevation view. These views give the reinforcement details of the base and the stub column (see, for example, Fig. 11.27). Alternatively, the tabular method may be used when a structure involves a large number of similar bases. In such a case, a typical base is drawn with reasonable dimensions and a schedule tabulating base details is included (see Fig. 11.28).

2 Combined bases for two or more columns. When two or more columns are close together and separate bases would overlap, a combined base can be used, as shown in Fig. 11.29. In combined bases, both top and bottom reinforcement bars are required. The main direction, in which steel bars are placed nearest to the base surface, could be any of the longitudinal or transverse directions.

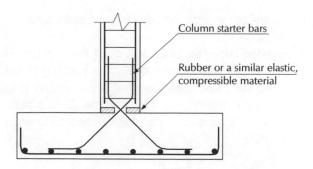

Figure 11.26   Hinged column/base connection

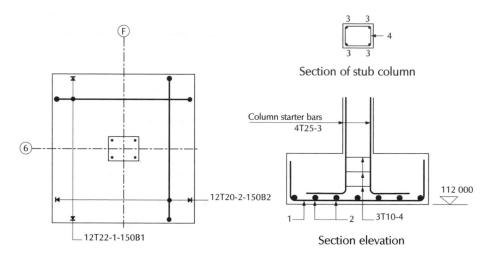

Figure 11.27   Reinforcement details of a concrete base

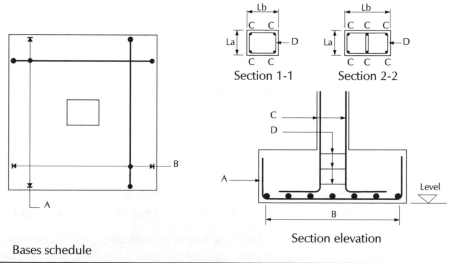

Bases schedule

Base	Number of bases	Section of column	Base dimensions (mm)			Base reinforcement		Column stub reinforcement		Level (m)
			X	Y	Z	A	B	Starters C	Links D	
A5,A6, C4,D6	4	1-1	1800	2200	450	16T20-1-150	13T20-2-150	4T22-3	3T8-4	110 000
A1,A2, C5	3	2-2	1800	2600	450	19T20-5-150	13T20-2-150	6T22-3	3T8-4	110 000

Figure 11.28   Details of concrete bases

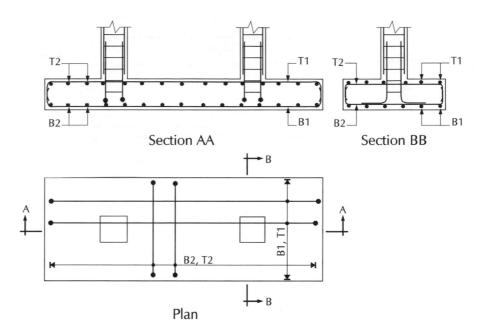

Figure 11.29 Combined concrete base

3 Wall foundations. Typical wall foundations, in which the wall is cast integral with the base, are shown in Fig. 11.30. Only bottom reinforcement bars are required, with the transverse direction being the main direction. If the wall is under horizontal forces in addition to the vertical forces, the projections of the base on the sides of the wall may not be equal.

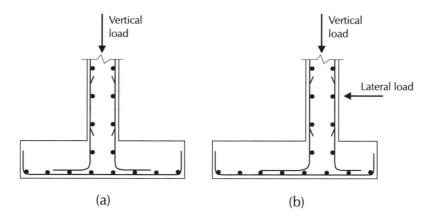

Figure 11.30 Sections through wall bases: (a) wall under a vertical load, (b) wall under vertical and lateral loads

11.8 Exercises

1. Draw a longitudinal section for each of the beams described below. Also draw a cross-section at every change of the beam reinforcement. Assume links of T8@200 mm in all beams:

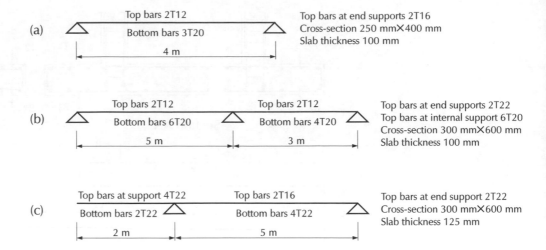

2. Draw plans and longitudinal and transverse sections for the slabs described below:
   (a) 6 m × 4 m slab with 200 mm thickness supported on four edge beams and reinforced by 30T12-200B1 and 20T12-200B2.
   (b) 5 m × 3.5 m slab with 200 mm thickness supported on two opposite beams 5 m long. The slab has reinforcement of 26T12-200B1 and 18T10-200B2.
   (c) A flat slab 200 mm thick, supported on columns with 6 m spacing in each direction. The width of the middle and column strips in both directions is 3 m. The slab is reinforced with 15T16-200B1, 20T16-150T1, 15T16-200B2 and 20T16-150T2 in the column strips and 15T12-200B1, 15T16-200T1, 15T12-200B2 and 15T16-200T2 in the middle strips.

3. Draw elevation and cross-section views of the walls described below to show their reinforcement details:
   (a) A perimeter wall of 250 mm constant thickness and 3.2 m height and supported on a strip base 1.4 m wide and 350 mm thick. The wall is reinforced with two layers each of T12 bars @150 mm in both directions. The base reinforcement is composed of T12-150B1 and 8T12-200B2.
   (b) A wall in a rectangular water tank has 300 mm thickness, 4 m width and 3.5 m height. The wall is reinforced with two layers each composed of T16 vertical bars @150 mm and T16 horizontal bars @200 mm.

4. Tabulate the details of the isolated bases described below. Choose the number and size of column starters and their links:
   (a) Bases A1, B2 and C5 have a size of 2.5 m × 1.5 m × 0.45 m and a column stub size of 0.8 m × 0.35 m. They are reinforced with 13T20-200B1 and 8T16-200B2.
   (b) Bases A3 and B1 have a size of 2.2 m × 1.7 m × 0.45 m and a column stub size of 0.5 m × 0.4 m. The base reinforcement is composed of 16T16-150B1 and 9T16-200B2.
   Also draw a plan and a cross-section of bases A1 and B1.

CHAPTER 12

Reinforced Concrete Structures

12.1 Introduction

In many small to medium sized structures, concrete is used as a construction material due to its efficiency and low cost compared with other materials. The advantages of concrete also include adequate durability, easy forming and high resistance to uplift forces in structures such as swimming pools.

In this chapter, common reinforced concrete construction systems are discussed with the aid of simple drawings showing how each is detailed and built safely. The discussion covers skeletal structures, shells, domes, precast structures and staircases. Comprehensive sets of drawings are also included and the reader is advised to study them carefully. This chapter must, however, be read in conjunction with Chapter 11 on concrete members and connections. Understanding the characteristics and detailing rules of concrete structures can not be achieved without a proper knowledge on how to detail components.

12.2 Skeletal structures

A skeletal reinforced concrete structure consists of a strong skeleton carrying all structural components including partition walls, floors and roof. It also carries all types of loading such as dead, live and wind loading. Concrete skeletal structures are similar to their steel equivalents, discussed before in Sec. 10.3. It was mentioned there that skeletal structures are either braced frame or unbraced frame structures, both composed of slabs, beams and columns. With concrete, a third type exists in which a structure is formed only of slabs and columns rigidly connected together, i.e. the flat slab system. Although different in constituents from unbraced beam/column structures, the flat slab system is an unbraced system due to its rigid column/slab joints (see Fig. 12.1). In flat slabs, beams may be used but only at the edges to strengthen the critical edge column/ slab joints.

In the following sections, the braced frame, the flat slab and the unbraced frame system are discussed.

Braced frame structures

Bracing of concrete frame structures is commonly done using shear walls (stiff vertical walls) or a structure's stiff core. This aspect of their design is not considered here as it needs a more specialized study.

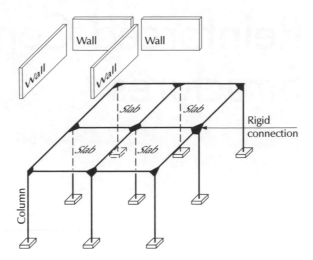

Figure 12.1   Flat slab system

LAYOUT OF BRACED FRAME STRUCTURES

Braced frame structures may have beams supported directly on columns such as the system shown in Fig. 12.2. The system may include secondary beams supported on main beams, which in turn are supported on the columns (see Fig. 12.3).

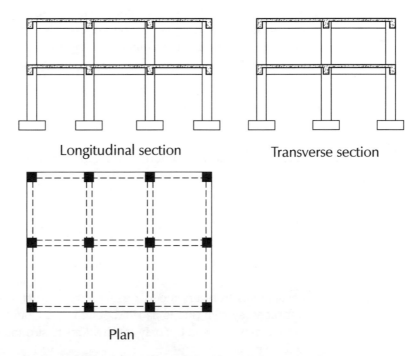

Figure 12.2   Layout views of a braced frame structure

BEAM/COLUMN AND SLAB/BEAM CONNECTIONS

Beam/column connections in a braced frame structure are typically designed to be hinged in that beams and columns are free to rotate independently. Thus, beams on one line are designed to be continuous (i.e. fixed to each other). Columns are also designed to be continuous through connections. Therefore, an ideal beam/column connection provides:

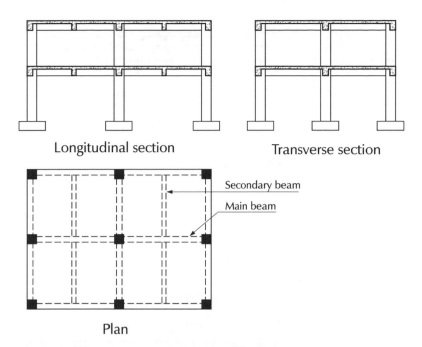

Figure 12.3 Layout views of a braced frame structure with main and secondary beams

(a) A *hinged* beam/column joint,
(b) A *fixed* beam/beam joint and
(c) A *fixed* column/column joint.

This can be achieved by controlling the arrangement of reinforcement bars at the connection (see for example Fig. 12.4). Notice that both the reinforcement of the beams and columns are continuous through the connection to develop fixed joints in (b) and (c) above, while the two groups are not interconnected to maintain a hinged beam/column joint.

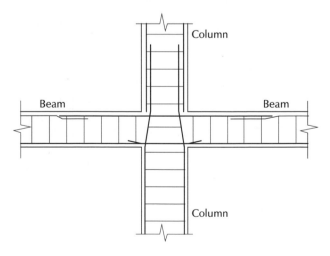

Figure 12.4 Typical beam/column connection

Similarly, slab/beam connections are also hinged. While slab reinforcement is continuous through the beam, there is no connection between the slab reinforcement and that of the beam (see Fig. 12.5).

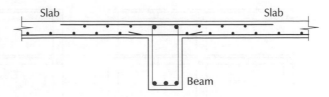

Figure 12.5   Typical slap/beam connection

REQUIRED DRAWINGS

The drawings required to describe the construction of a reinforced concrete braced frame structure are listed below. Refer to the typical example presented in Figs 12.6 to 12.10.

1   A layout plan, normally on a rectangular grid, showing the position and size of bases. The reinforcement details of bases are scheduled on the same drawing (see Fig. 12.6).

2   A layout plan giving the positions and sizes of columns. A schedule of column reinforcement is included in the same drawing (see Fig. 12.7).

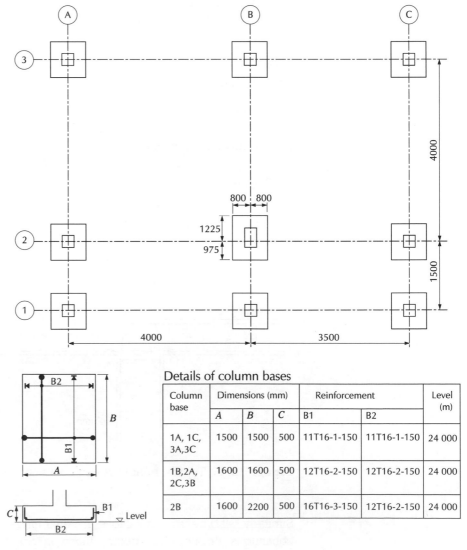

Details of column bases

Column base	Dimensions (mm)			Reinforcement		Level (m)
	A	B	C	B1	B2	
1A, 1C, 3A,3C	1500	1500	500	11T16-1-150	11T16-1-150	24 000
1B,2A, 2C,3B	1600	1600	500	12T16-2-150	12T16-2-150	24 000
2B	1600	2200	500	16T16-3-150	12T16-2-150	24 000

Figure 12.6   Layout plan and details of foundations

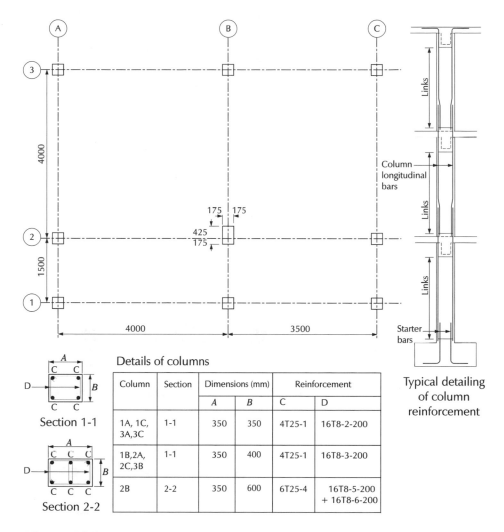

Figure 12.7 Layout plan and details of columns

Details of columns

Column	Section	Dimensions (mm)		Reinforcement	
		A	B	C	D
1A, 1C, 3A,3C	1-1	350	350	4T25-1	16T8-2-200
1B,2A, 2C,3B	1-1	350	400	4T25-1	16T8-3-200
2B	2-2	350	600	6T25-4	16T8-5-200 + 16T8-6-200

3 A layout elevation showing the number of floors and levels of floors and bases.

4 A small scale plan of every floor showing slab thickness and reinforcement, beam serial numbers and sizes (see Figs 12.8 and 12.9).

5 Large scale longitudinal and cross-sections of beams giving all reinforcement details (see Fig. 12.10).

Figures 12.6 to 12.10 present a set of drawings of a simple beam/column structure. The same principles apply to the detailing of larger structures.

Flat slab structures

Flat slab structures are also skeletal, but unlike braced frame systems, they do not have internal beams and the connections between slabs and columns are rigid. To achieve rigidity, slab/column connections are usually prepared with a column head and/or a slab drop (see Fig. 12.11).

A general layout view of a flat slab structure with column heads and slab drops is shown in Fig. 12.12. Notice that the column heads of edge and corner columns are incomplete.

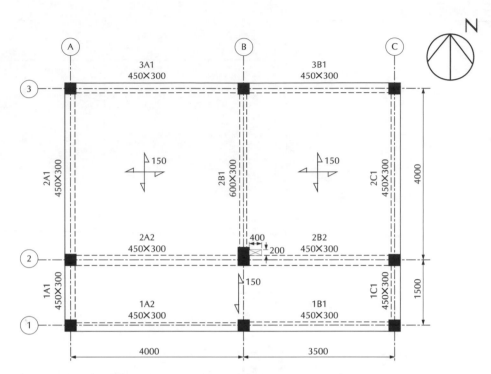

Figure 12.8   Slab and beam details on a small-scale floor plan

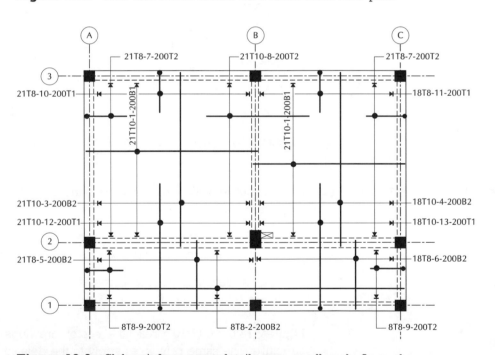

Figure 12.9   Slab reinforcement details on a small-scale floor plan

The construction drawings of flat slab structures are similar to those of braced frame structures. The main difference is in the detailing of the slab reinforcement. An example on flat slab reinforcement detailing is presented in Fig. 11.17.

Unbraced frame structures   The discussion in this section covers only planar frame structures. However, the same principles in connection design and overall detailing also apply to three-dimensional frame structures.

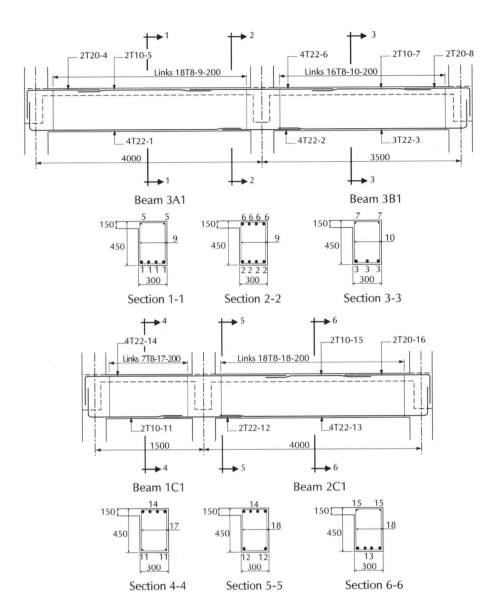

Figure 12.10 Large scale views and cross-sections of floor beams

BEAM/COLUMN AND SLAB/BEAM CONNECTIONS

An unbraced frame structure is made up of beams and columns that are connected to one another at rigid connections which do not allow relative rotation to occur between the ends of the connected members. Many unbraced frame structures resemble simpler braced frame structures in appearance, but are radically different in behaviour due to the joint rigidity. The main difference between the two systems is in the connection detailing, and mainly that of the reinforcement bars (see Fig. 12.13). In braced frame structures, each of the beam and column reinforcement is continuous through the connection, but separate from each other. In contrast, there is a continuity of reinforcement bars of both the beam and column in unbraced frame structures. On the other hand, slab/beam connections are identical to those in braced frame structures (see Fig. 12.5).

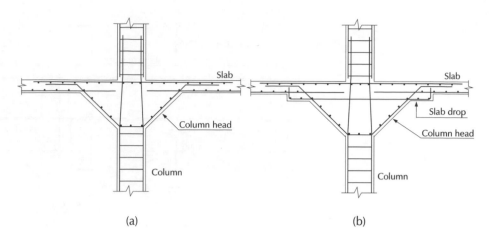

Figure 12.11   Typical slab/column connections in a flat slab structure: (a) with a column head, (b) with a column head and a slab drop

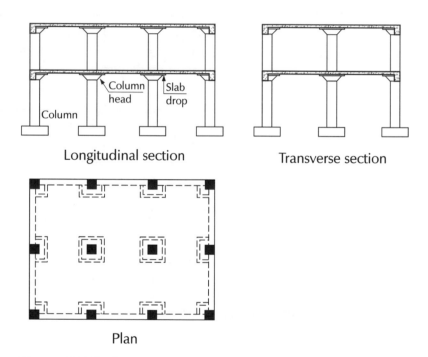

Figure 12.12   Layout views of a flat slab structure

OVERALL FORMS OF PLANAR FRAMES

Various forms of RC planar frames are in use including the rectangular (also called portal), pitched-roof, multi-bay, multi-storey and multi-bay/storey frames (see Fig. 12.14).

In planar frame structures, frames are normally constructed in parallel lines and connected by slabs and beams at floor and roof levels. Also, beams within the height of columns are sometimes used. As an example, Fig. 12.15 shows a rectangular planar frame carrying a slab and five transverse beams. Like metallic planar frames, a concrete planar frame structure requires a bracing system which is provided in this case by the slabs and secondary beams acting to stabilize the building against lateral loads. Alternatively, a set of shear walls may be built, especially in tall structures.

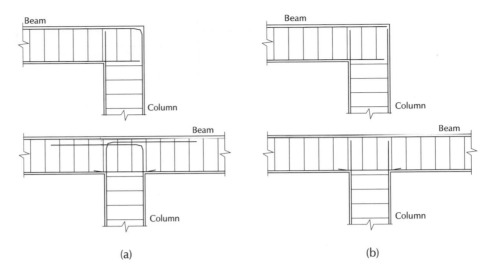

Beam

Column

Beam

Column

Beam

Column

Beam

Column

(a)   (b)

Figure 12.13 The use of reinforcement bars to develop rigid and hinged connections: (a) rigid connections in unbraced frame structures, (b) hinged connections in braced frame structures

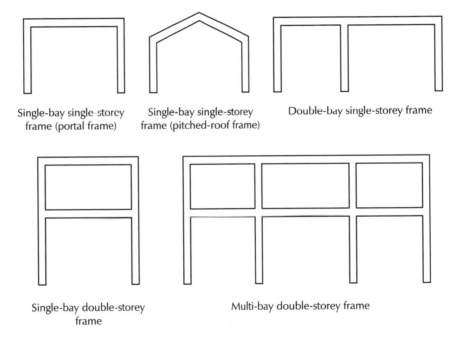

Single-bay single-storey frame (portal frame)

Single-bay single-storey frame (pitched-roof frame)

Double-bay single-storey frame

Single-bay double-storey frame

Multi-bay double-storey frame

Figure 12.14 Planar frames

For frames with inclined main beams, such as the pitched-roof frame, slabs might also be inclined, as shown in Fig. 12.16.

THE SUPPORTS

The connections between columns and bases in planar frame structures may be fixed or hinged according to the design requirements. Examples of a fixed and a hinged connection are shown in Figs. 11.25 and 11.26 respectively.

REQUIRED DRAWINGS

The drawings required for the construction of planar frame structures are listed below. Also refer to the example given in Figs 12.17 and 12.18.

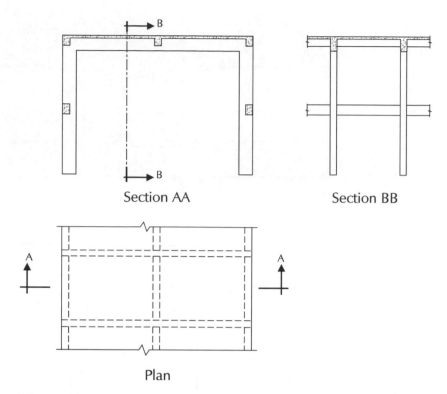

Section AA   Section BB

Plan

Figure 12.15   Views of a rectangular (portal) frame structure

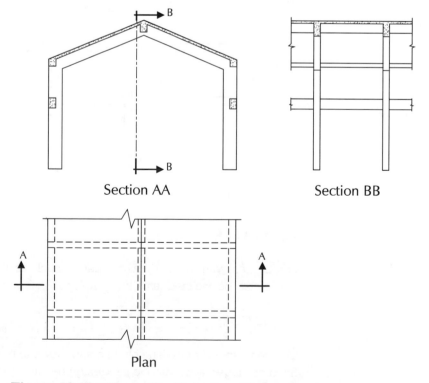

Section AA   Section BB

Plan

Figure 12.16   Views of a pitched-roof frame structure

1   A layout plan giving the position and size of bases. A schedule of the base reinforcement details is included on the same drawing. This drawing is similar to that of a braced frame structure (see, for example, Fig. 12.6).

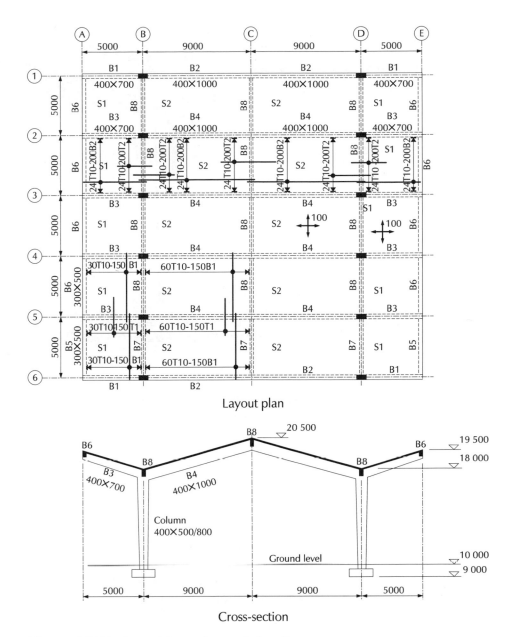

Figure 12.17 Layout floor plan and elevation

2 A layout plan showing for every floor the slab thickness and reinforcement and beam serial numbers and sizes (see Fig. 12.17).

3 A layout elevation showing the number of floors and levels of floors and bases (see Fig. 12.17).

4 Large scale detailed sectional views of main frames giving the reinforcement details (see Fig. 12.18). If symmetric, only half of the frame may be detailed.

5 Large scale detailed views of the secondary beams.

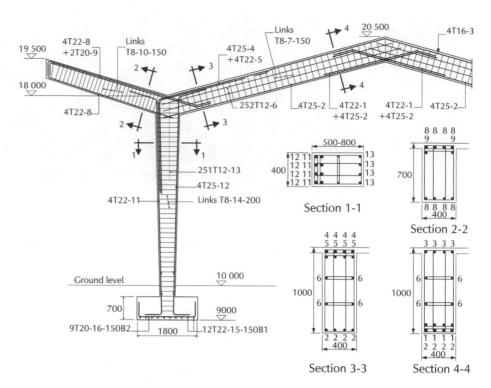

Figure 12.18 Large-scale detailed sectional view of main frame with reinforcement details

12.3 Reinforced concrete shell structures

COMMON SHELL FORMS

A shell is a thin, rigid, three-dimensional structural form that encloses a volume by a curved surface. Shell forms include:

- Rotational surfaces generated by the rotation of a curve about an axis (e.g. spherical and parabolic surfaces) and
- Translational surfaces generated by sliding of one plane curve along another plane curve (e.g. cylindrical and elliptic paraboloid surfaces) (see Fig. 12.19).

The most common forms are the spherical and cylindrical shells.

Shell structures carry the applied loads by the development of in-plane stresses (i.e. tension, compression and shear stresses acting within the thickness of the shell). As a consequence of this load-carrying feature, shell structures can be very thin and still span large distances, provided that a suitable material such as reinforced concrete is used. Masonry shells, for instance, are considerably thicker than concrete shells.

CYLINDRICAL SHELLS

Concrete cylindrical shells commonly have stiff diaphragms at the ends and the internal supports, if any are used (see Fig. 12.20). The diaphragms prevent the

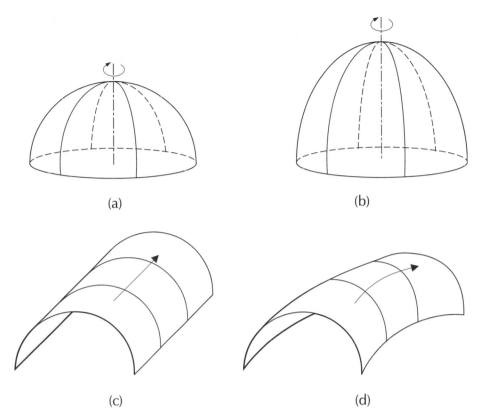

Figure 12.19 Common form of shells. (a) Spherical shell (dome) generated by rotation of an arc about an axis. (b) Parabolic shell (dome) generated by rotation of a parabola about an axis. (c) Cylindrical shell generated by translation of an arc along a straight line. (d) Elliptic paraboloid shell generated by translation of a curve along another plane curve.

distortion of the shell and mainly its tendency to flatten out. Longitudinal beams are used to provide sufficient reinforcement for the bottom parts of the shell and thus prevent their cracking.

Due to the thinness of shells and the fact that they carry mainly in-plane stresses, only one layer of reinforcement is used at mid-thickness of the shell. This layer is strengthened by another near the longitudinal beams so that any local moment resulting from the connection to the beams can be resisted (see Fig. 12.21).

DOMES

Domes, on the other hand, are spherical shells (parabolic domes are rarely used). They are ideal for covering circular or square areas. Domes are mainly used in buildings such as churches, mosques, sports centres, exhibition halls, airport terminals, etc. A dome should have a circular ring beam at its bottom edge to carry the tension hoop forces which exist there, and without which cracking would occur (see Fig. 12.22).

Notice from Fig. 12.22 that domes, like cylindrical shells, have only one reinforcement layer at mid-thickness. When concentrated forces are applied on a dome, e.g. the weight of a lantern, the loaded part must be strengthened by increasing its thickness and reinforcement.

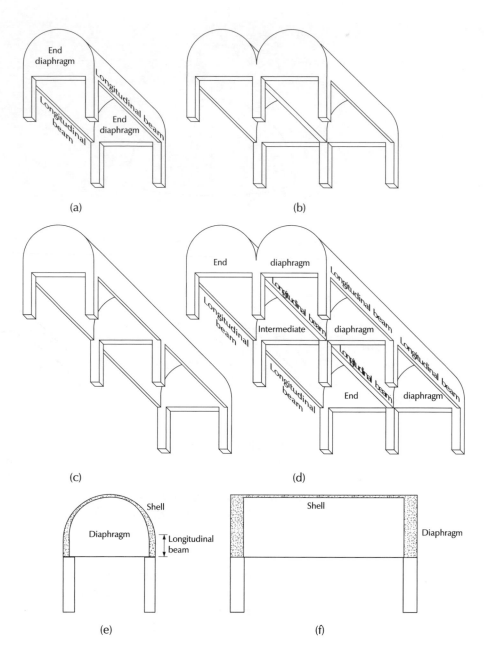

Figure 12.20 Reinforced concrete cylindrical shells. (a) One-bay shell over one span. (b) Two-bay shell over one span. (c) One-bay shell over two spans. (d) Two-bay shell over two spans. (e) Cross-section of a one-bay shell. (f) Longitudinal section of a single-span shell

REQUIRED DRAWINGS

The drawings required for the construction of a concrete shell structure are listed below. Refer to the examples presented in Figs 12.23 and 12.24 of a cylindrical shell and a spherical dome respectively.

1 Layout orthogonal views showing shell dimensions and levels, column positions, base levels, etc.

2 A large scale cross-sectional view of the shell showing the reinforcement detailing of both the shell and the longitudinal beams (or the ring beam of a dome).

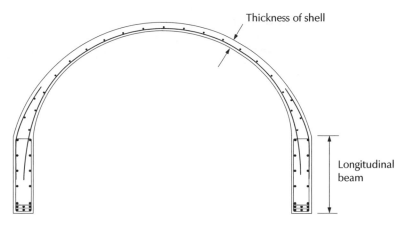

Figure 12.21   Cross-section of a one-bay shell

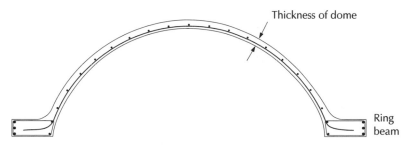

Figure 12.22   Cross-section of a dome

3   A large scale plan of the shell showing the reinforcement details. In some cases it might be appropriate to develop the cylindrical shell to show the distribution of shell bars along with those of the longitudinal beams (see Fig. 12.25). In the case of a dome, the plan is drawn to the actual size (see, for example, Fig. 12.24).

4   Longitudinal and cross-sectional views of the diaphragms of cylindrical shells showing their reinforcement and the connection with the columns.

Figure 12.25 is presented here to explain how developed plan views are drawn for cylindrical shells. The diagonal bars shown are used, if necessary, to resist shear stresses.

12.4 Stairs and staircases

Stairs are an essential part of every building since they lead from one floor to another and connect different levels. Figure 12.26 shows three types of staircase and explains the basic terminology of stairs and staircases. The direction of rise can be shown on plan views of stairs by numbering the stairs in an ascending order and using an open-head arrow pointing towards the higher level. The maximum allowable values for the rise and going of stairs are given in Table 12.1 for staircases in different situations.

In staircases, each flight together with its end landings act as a slab spanning in one way between beams at two different levels. Staircases may be supported in various ways, and their reinforcement detailing changes accordingly. Figure 12.27 shows the common cases of stair slabs supported on two, three or four

Table 12.1

Staircase type	Maximum rise (mm)	Maximum going (mm)
Lightly used staircases	220	220
Commonly used staircases	190	240
Staircases of public and institutional buildings	180	250

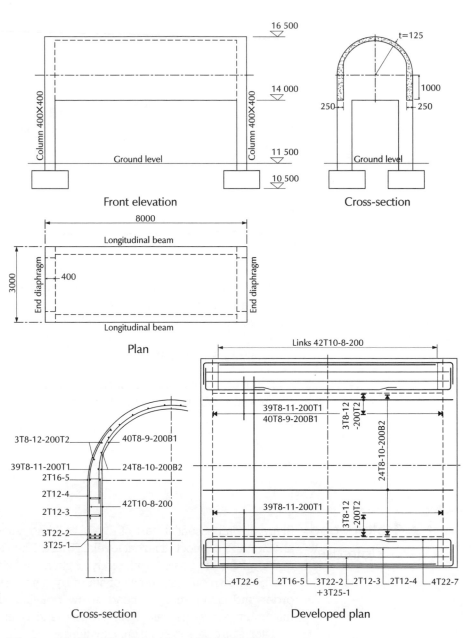

Figure 12.23　Detailing of a cylindrical shell

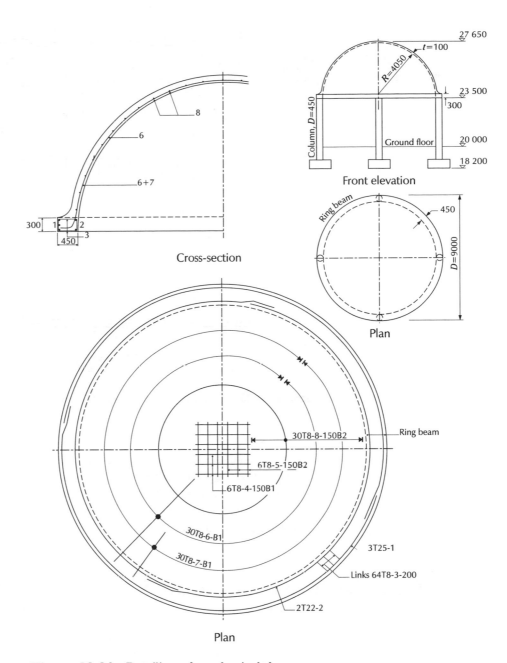

Figure 12.24 Detailing of a spherical dome

beams. Stair slabs may also be supported on walls. However, in most circumstances, the reinforcement detailing is almost the same whether the slab is supported on a wall or a beam.

There are other types of stairs such as stairs cantilevered from a side beam (or a side wall), free-spanning spiral stairs and spiral stairs cantilevered from a central column. They are not included in this book as their use is very limited.

The drawings required for the construction of staircases include:

1 A plan view showing the number of flights per storey, the number of stairs per flight, the direction of rise and the beams supporting the staircase. The plan also includes all dimensions required to set out the concrete work.

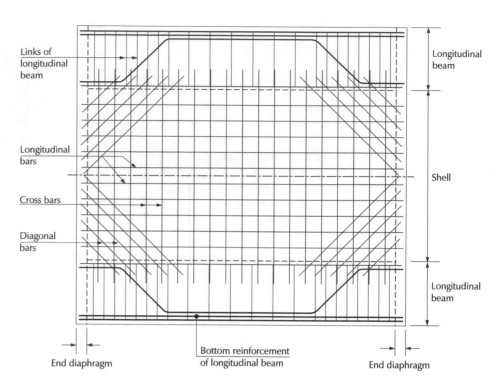

Figure 12.25   Reinforcement details of a cylindrical shell on a developed plan

2   Longitudinal sectional views of the flights showing the distribution of longitudinal and transverse reinforcement bars, thickness of slab and connection with the staircase beams. The details of the supporting beams and the surrounding slabs are not shown as they can be obtained from the drawings detailing the floors.

12.5 Exercises

1   What is the difference between braced frame and unbraced frame structures?

2   For the structure whose roof plan is given in Fig. 12.28 draw detailed orthogonal views to show column, base, slab and beam details. The structure has one storey of 3 m high above ground level.

   (a)   Columns A1, A2 and A3 are 0.3 m × 0.3 m in section and reinforced with 4T16 and T8 links @200 mm.

   (b)   Columns B1, B2, B3, D1, D2 and D3 are 0.3 m × 0.5 m in section and reinforced with 6T16 and 2T8 links @200 mm.

   (c)   Bases A1, A2 and A3 are 1.2 m × 1.2 m × 0.35 m in size and reinforced with 9T12-150B1 and 9T12-150B2.

   (d)   Bases B1, B2, B3, D1, D2 and D3 are 1.2 m × 1.6 m × 0.35 m in size and reinforced with 12T16-150B1 and 9T12-150B2.

   (e)   All beams are 0.3 m × 0.5 m in section and all slabs are 0.1m thick.

Assume beam and slab reinforcement.

3   A portal frame structure consists of seven identical frames, each of 12 m span and 6 m height above bases. Frame spacing is 4 m. The frames are connected by a 100 mm thick top slab and three 0.3 m × 0.6 m cross beams at top corners and mid-span. Draw layout views of the structure showing all

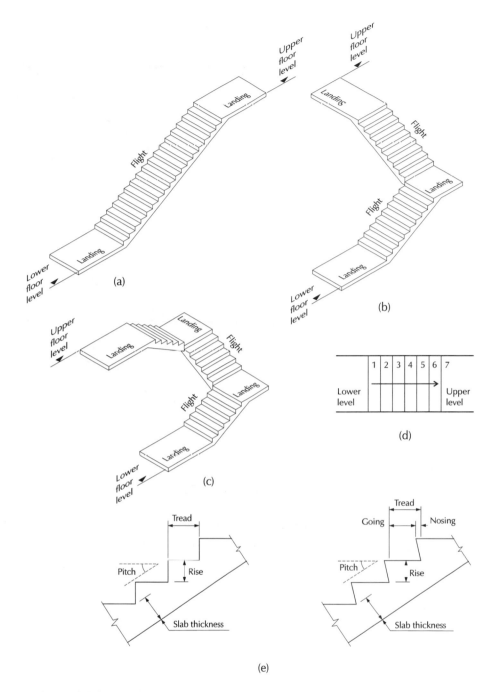

Figure 12.26 Reinforced concrete staircases: (a) one-flight staircase, (b) two-flight staircase, (c) three-flight staircase, (d) plan view of a staircase, (e) cross-section through stairs

member dimensions. Also draw large scale sectional views to detail the reinforcement of the frames and cross beams. Figure 12.29 gives the reinforcement magnitudes of a typical frame.

4   Draw a plan and cross-section of the single-bay single-span shell whose span is 9 m and section is as given in Fig. 12.30.

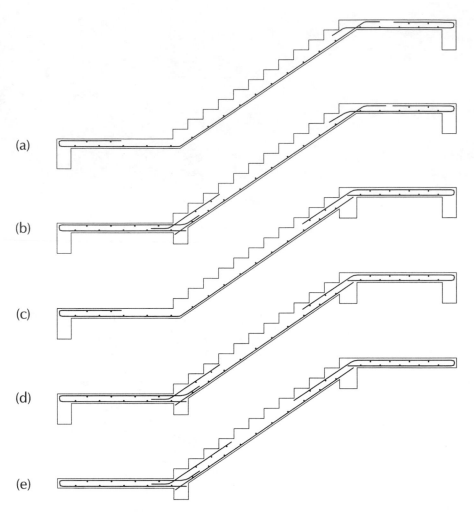

Figure 12.27   Reinforcement details of stair slabs. Slab supported on (a) two edge beams, (b) and (c) three beams, (d) four beams, (e) two internal beams

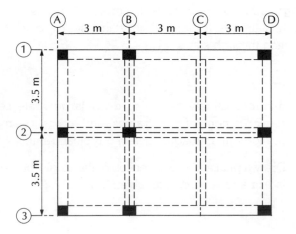

Figure 12.28

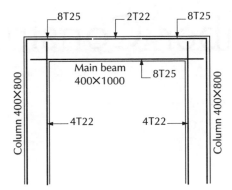

Figure 12.29

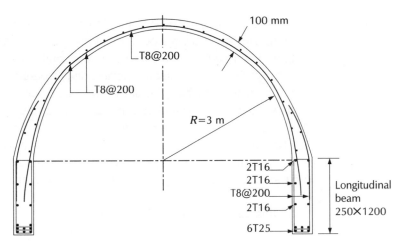

Figure 12.30

CHAPTER 13 Timber Construction

13.1 Introduction

Timber was man's first construction material, and still it is as important today as ever. Timber is sawn wood, a material obtained from trees, and primarily from their main stems. Ancients used it to build their bridges which were made simply of one tree trunk supported on the banks of a river. Timber was used in building huts, making furniture and manufacturing tools. Nowadays, timber is used in footbridges and as roofs in modern houses, bungalows, exhibition halls, sports centres, etc. There are more timber framed dwellings built in the world each year than houses built using any other material. Newton's footbridge on River Cam in Cambridge is one of many examples of the efficiency and long life span of timber when properly treated and maintained.

In this chapter, the different types of timber members and fasteners are presented, followed by a discussion on timber trusses and footbridges.

13.2 Timber specification

When specifying timber for a project, the following should be clearly stated:

1 Strength grade, for timber has various grades of strength, each being suitable for a specific usage.

2 Finished cross-sectional size and tolerance. Since producing a timber section with exact dimensions is almost impossible, the acceptable tolerance must be specified. This might dictate the method of cutting and finishing of timber members.

3 Length of a piece, or otherwise reference should be made to the relevant drawings.

13.3 Fasteners

Timber members may be fastened to each other or to concrete or steel members using either glue or a mechanical fastener.

Glued joints

The glue used in timber joints should be water-proof and boil-proof to a degree that depends mainly on the exposure of timber to the elements. Some typical glued joints are shown in Fig. 13.1. Perhaps the most efficient glued joint is the finger joint shown in Fig. 13.1d. Finger joints are self-locating joints formed by machining a number of similar tapered symmetrical fingers in the ends of timber members which are then glued and interlocked under pressure.

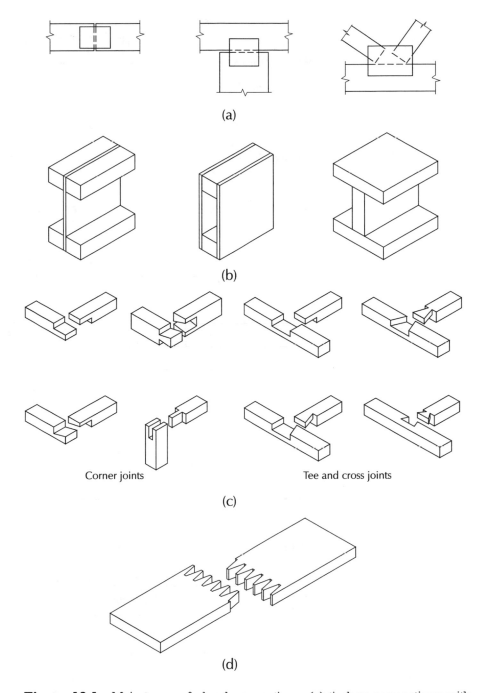

Figure 13.1 Main types of glued connections: (a) timber connections with splice plates, (b) built-up beam members, (c) common simple timber joints, (d) finger joint

Mechanical fasteners

Several types of mechanical fasteners may be used in timber joints, the most common of which are nails, screws, bolts, split-ring connector units and steel hangers.

NAILS

Various sizes of nail are available on the market. Ordinary nails, shown in Fig. 13.2a are round and range in diameter between 2.65 and 8 mm, and in length between 15 and 150 mm. This type of nail should only be used while perpendicular to the direction of load (see Fig. 13.3). When withdrawal loads are

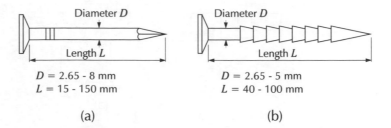

$D = 2.65 - 8$ mm
$L = 15 - 150$ mm

$D = 2.65 - 5$ mm
$L = 40 - 100$ mm

(a)   (b)

Figure 13.2   Nails: (a) ordinary nail, (b) annular-ringed shank nail

expected, special nails such as the annular-ringed shank nails are used (see Fig. 13.2b).

Figure 13.3 shows some examples of typical nailed joints. A minimum of two nails is normally acceptable in a structural joint.

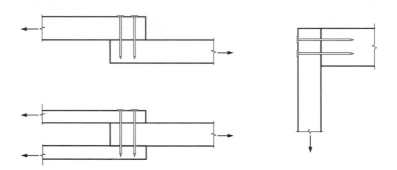

Figure 13.3   Typical examples of nailed joints

SCREWS

Screws are used principally for fixings that require a resistance to withdrawal greater than that provided by nails. Screws may have slotted countersunk heads, round heads or square heads, all shown in Fig. 13.4. Screwed joints are similar to the nailed joints shown in Fig. 13.3 in that a minimum of two screws per joint is acceptable, and screws must be used while perpendicular to the direction of load.

BOLTS

The bolts used in timber joints are made of ordinary mild steel, black or galvanized. Large washers are usually used under the head and nut of a bolt to reduce the bearing stress on the timber. The bolts available range between 8 and 36 mm in diameter, and between 35 and 300 mm in length. In bolted joints, the forces may be parallel, inclined or perpendicular to the grains of a cross member (see Fig. 13.5). Notice that bolted joints may connect two, three or sometimes more members.

Spacing of bolts should be in accordance with Fig. 13.6 with due regard to the directions of load and timber grains. When the applied load is inclined to the grains, the adopted spacings should equal the greater of the recommended spacings for the parallel and perpendicular cases.

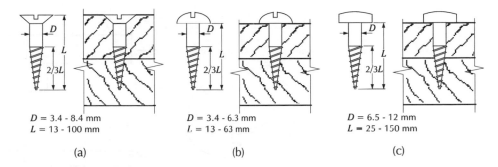

D = 3.4 - 8.4 mm
L = 13 - 100 mm
(a)

D = 3.4 - 6.3 mm
L = 13 - 63 mm
(b)

D = 6.5 - 12 mm
L = 25 - 150 mm
(c)

Figure 13.4 Main types of screw: (a) slotted countersunk head screw, (b) round head screw, (c) square head screw

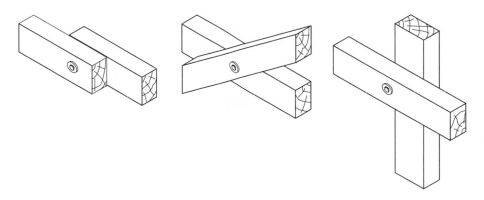

Figure 13.5 Typical bolted joints

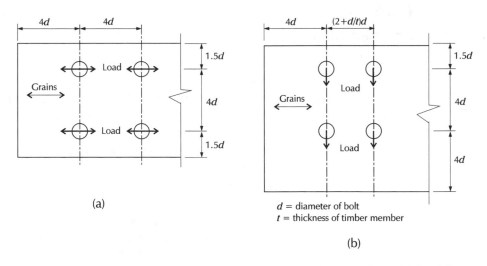

d = diameter of bolt
t = thickness of timber member

(b)

Figure 13.6 Minimum spacing of bolts in bolted timber joints: (a) load is parallel to grains, (b) load is perpendicular to grains

SPLIT-RING CONNECTOR UNITS

A split ring is a steel ring embedded in grooves made specially in two timber members to maintain a strong connection between them. Figure 13.7 shows a typical split ring and how it is used to make a two-member and a three-member joint.

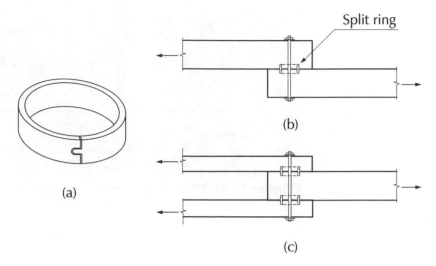

Figure 13.7 Split-ring timber joints: (a) split ring, (b) two-member joint, (c) three member joint

STEEL HANGERS

Steel hangers are used to fix timber beams to steel, concrete, masonry or timber main beams or columns. Several types of hanger are available for use for different purposes. Figure 13.8 shows some of the commonly used hangers and how they are used to make a joint.

13.4 Timber structural members

Timber structures can be subdivided into basic elements such as a decking, beams, columns and bases. These elements are discussed in the following sections.

Timber decking

Solid timber decking is mainly used in roofs where the decking spans between timber beams giving a solid, permanent roof deck and a ceiling with a pleasant appearance. Decking is normally made of timber boards of the cross-sections shown in Fig. 13.9, which also shows how boards span on cross beams. Spanning the boards over single spans, double spans or single/double spans is a common practice.

In footbridges where the appearance of decking is not critical, boards of a rectangular cross-section are used while extending longitudinally or transversely. A typical example of footbridge decking is presented later in Fig. 13.25.

Timber beams

BEAM CROSS-SECTIONAL SHAPES

There are several cross-sectional shapes of timber beams in use, the more common of which are shown in Fig. 13.10. Beams may be rectangular and made from one timber piece or composed of a number of pieces, usually glued together.

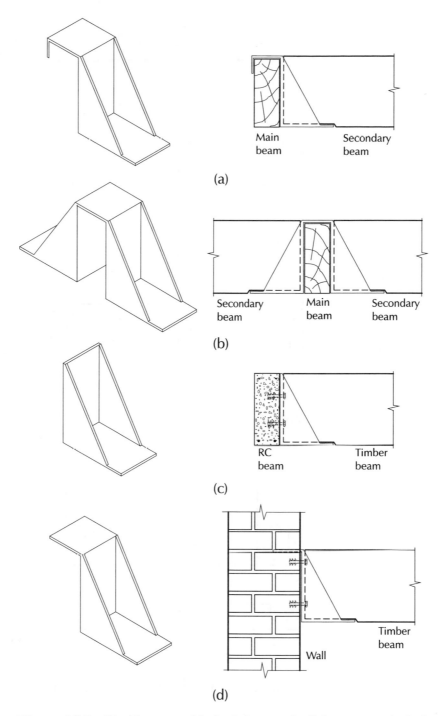

Figure 13.8 Steel hangers: (a) single hanger for fixing on a main timber beam, (b) double hanger for fixing on a main timber beam, (c) single hanger for fixing on an RC beam, (d) single hanger for fixing to a masonry wall

BEAM JOINTS

When a timber piece of a composite beam is too short to cover the whole length of the beam, a glued or a spliced joint is used as shown in Fig. 13.11. Notice that in spliced joints the thickness of each of the splice plates is the same as that of the original plate.

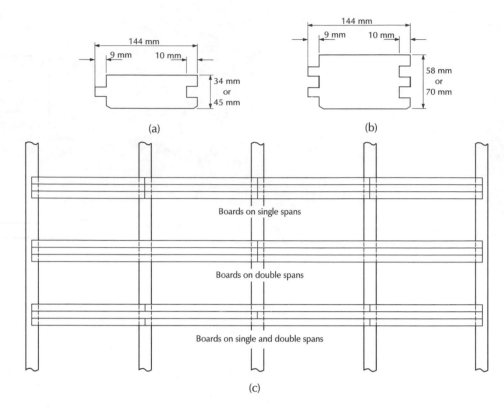

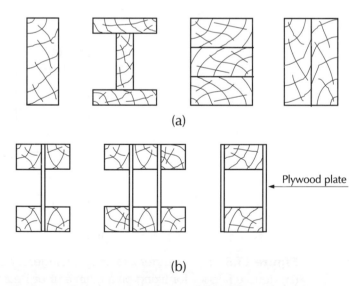

Figure 13.9 Timber decking: (a) cross-section of a shallow board, (b) cross-section of a deep board, (c) layout of boards on cross beams

Figure 13.10 Cross-sections of typical timber beams: (a) beams made from one or more pieces, (b) composite beams involving plywood plates

BEAM SUPPORTS

Beams may be supported on columns in various ways:

1 Beams may be continuous through the connection while supported on the column side (Fig. 13.12a).

2 Beams may be simply supported on the column side (Fig. 13.12b).

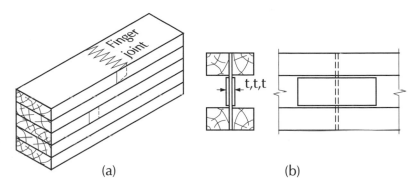

Figure 13.11 (a) Glued and (b) spliced joints

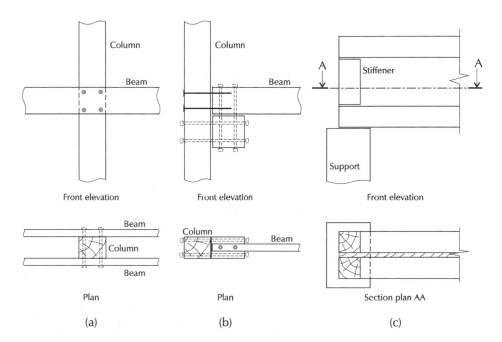

Figure 13.12 Beam supports

3 Beams may be resting on top of the support (Fig. 13.12c).

Web stiffeners might be needed for beams with plywood webs, especially at the supports.

COMPOSITE DOUBLE-SKIN PANELS

In some cases, it might be appropriate to use composite floors built of beams between two skins of plywood plates and acting compositely with them (see Fig. 13.13). This form of construction is denoted as a double-skin panel. The spaces between the beams can be used for ventilation, electrical wiring or even plumbing.

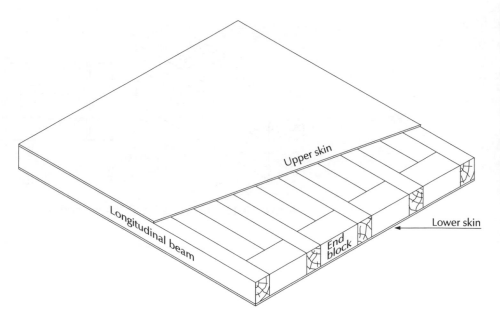

Figure 13.13   A composite floor unit (double-skin panel)

Timber columns

COLUMN CROSS-SECTIONAL SHAPES

The common sections of timber columns are similar to beam sections. Columns may be made of one timber piece or composite of a number of pieces to provide adequate strength and stiffness against axial loading as well as any possible lateral loading. Figure 13.14 shows some typical cross-sections of timber columns.

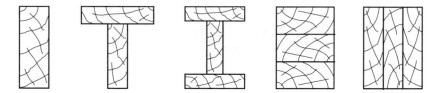

Figure 13.14   Cross-sections of timber columns

BUCKLING OF COLUMNS

To avoid buckling of long timber columns, they are usually tied laterally to nearby brick or concrete walls where appropriate (see Fig. 13.15). This adds to the stability of the column and increases its load-carrying strength.

Beam/column connections of timber structures have been discussed in the previous subsection and examples are given in Fig. 13.12.

Bases   The base of a column usually takes the form of a steel shoe bearing on a concrete footing (see Fig. 13.16). An adequate adhesive is used to fix the column into the steel shoe and therefore avoid any possible uplifting of the column.

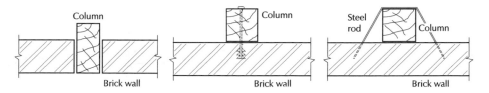

Figure 13.15 Timber columns tied to prevent buckling

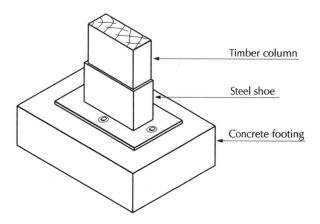

Figure 13.16 Footing of a timber column

In some cases when the soil condition allows, it might be appropriate to prepare the column with a steel base with a sharp end, and then drive it into the ground as a pile.

13.5 Timber structures

The most common structural uses of timber is in trusses and footbridges; these are discussed in the following sections.

Timber trusses

TRUSS OVERALL FORMS

Timber trusses are mainly used as roofs of houses and other buildings. The choice of a truss form is usually influenced by architectural considerations, type and size of roof material, support conditions, span and economy. In most cases, the truss form is chosen from three basic truss types: the pitched-roof, the bowstring and the parallel-chord trusses all shown in Fig. 13.17.

TRUSS JOINTS WITH GUSSET PLATES

Truss joints can be formed using mild steel gusset plates similar to those used with metallic trusses. Gusset plates are usually fitted on the outside faces of single-chord trusses, as shown in Fig. 13.18a. Normally, two gusset plates are used, one on each side of the joint. When it is not visually acceptable to have the gusset plates exposed to view, double-chord trusses are considered, in which the gusset plates are hidden between the two chords, see Fig. 13.18b.

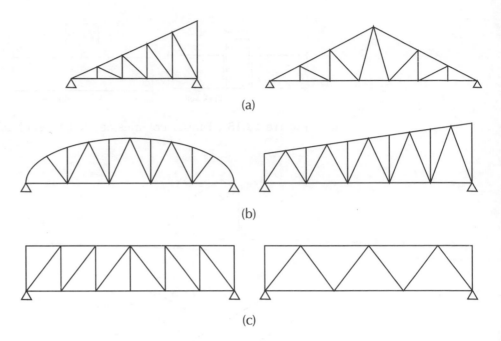

Figure 13.17   Common forms of timber trusses: (a) pitched-roof, (b) bowstring and (c) parallel-chord trusses

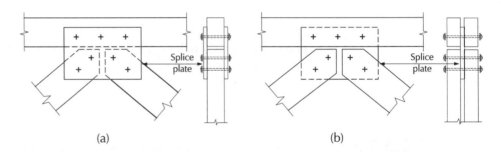

Figure 13.18   Spliced truss joints: (a) single-chord and (b) double-chord truss joints

BOLTED JOINTS OF LIGHTLY-LOADED TRUSSES

For lightly loaded trusses, or trusses on short spans, simpler multi-member joints may be used, in each of which one or two bolts are used. The resulting joint suffers from some eccentricity which must be considered in design. An example truss is shown in Fig. 13.19 where two of its joints are also sketched.

TENSION MEMBERS OF TRUSSES

In timber trusses when some members carry only tension forces under all possible loading cases, these members may be replaced by steel rods (see, for example, Fig. 13.20).

END JOINT OF PITCHED-ROOF TRUSS

The end joint of the pitched-roof truss shown in Fig. 13.20 is very critical. Relative slip between members or a local failure may cause an overall collapse of the

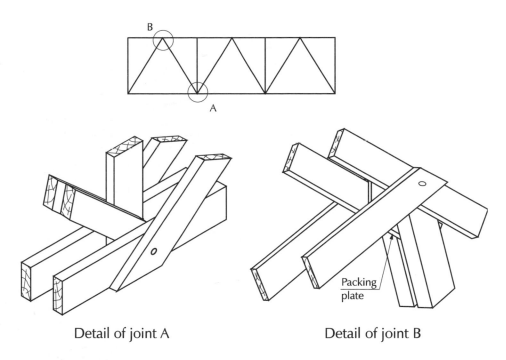

Detail of joint A   Detail of joint B

Figure 13.19 A lightly loaded truss and details of two of its joints

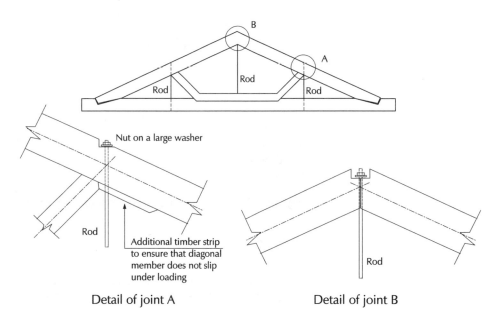

Detail of joint A   Detail of joint B

Figure 13.20 A pitched-roof timber truss

truss or at least a severe distortion. Figure 13.21 shows a typical form of this joint. The support shown in Fig. 13.21 may be a concrete or masonry wall or column over which the truss is resting, or it may be a timber column, properly fixed to the truss, usually using connecting steel plates on the sides of the joint.

THE BRACING SYSTEM

Similar to steel trusses, a bracing system is a necessity for timber trusses in order to provide lateral stability. The bracing system should adequately connect all trusses together and form a structural system able to transmit the lateral forces to the ground supports (see, for example, Fig. 13.22).

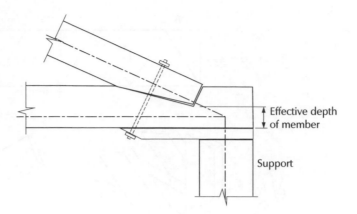

Figure 13.21   Details of the end connection of a pitched-roof truss

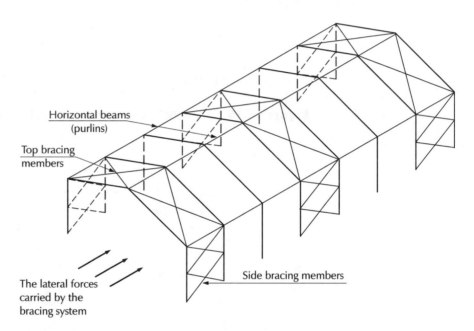

Figure 13.22   Bracing system of a timber truss structure

REQUIRED DRAWINGS

The drawings required to describe the construction of timber trusses are listed below. Refer to the example presented in Figs 13.23 and 13.24.

1   A layout plan showing the positions and sizes of columns and bases. The reinforcement of the concrete bases could be detailed on the same drawing (see Fig. 13.23).

2   Layout plan and elevation of the structure showing trusses, purlins and bracing system with all member dimensions.

3   A layout elevation of a typical truss showing dimensions and member joints (see Fig. 13.24). All members are also detailed on the same drawing.

4   Large scale detailing of complicated joints and packing and gusset plates.

Figures 13.23 and 13.24 present the detailing of a simple pitched-roof timber truss.

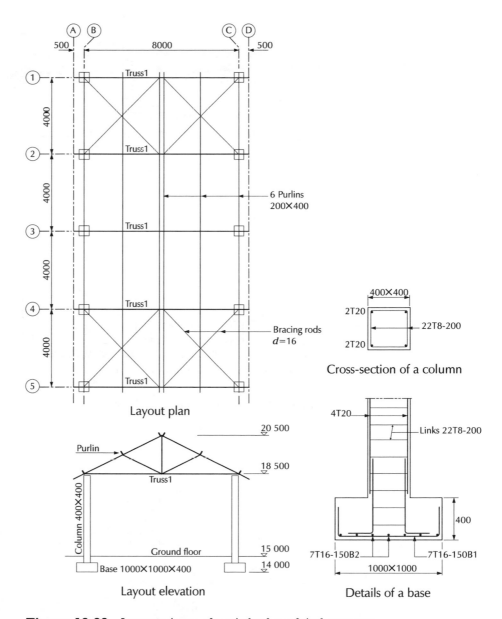

Figure 13.23 Layout views of a pitched-roof timber truss

Timber footbridges   Timber is a common material in the construction of small- to medium-span footbridges. Timber footbridges normally consist of longitudinal or transverse decking boards supported on longitudinal timber beams, or sometimes steel beams. Figure 13.25a shows four common forms of timber footbridges with their handrails and toe boards. The main beams of the bridge may be supported on a reinforced concrete abutment or on vertical timber columns as shown in Fig. 13.25b.

Figure 13.26 presents a typical example of the detailing of a timber footbridge. The bridge has 7.1 m length and 1.5 m width, and is supported on end reinforced concrete walls. It is detailed using three orthogonal views. Since the timber pieces are all prismatic, no member detailing is necessary. In addition, the joints are simple with one or two bolts at each joint.

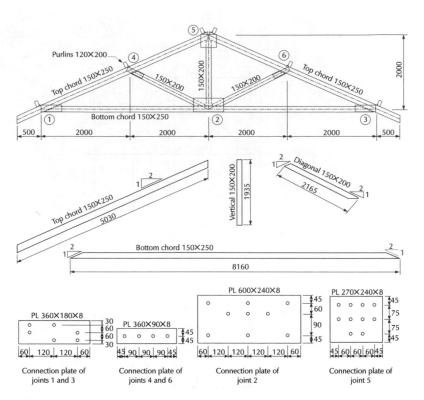

Figure 13.24   Detailing of a pitch-roof timber truss

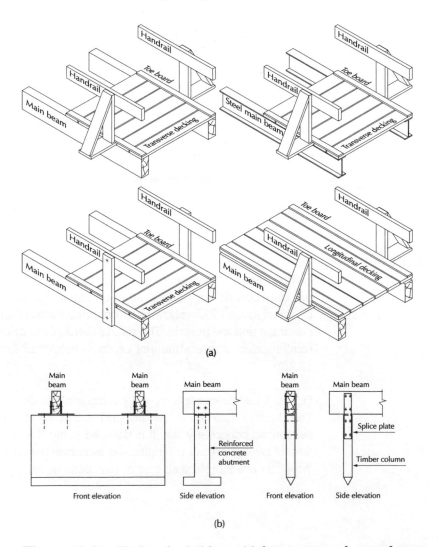

Figure 13.25   Timber footbridges: (a) four common forms of construction (b) main beams supported on concrete abutments or timber columns

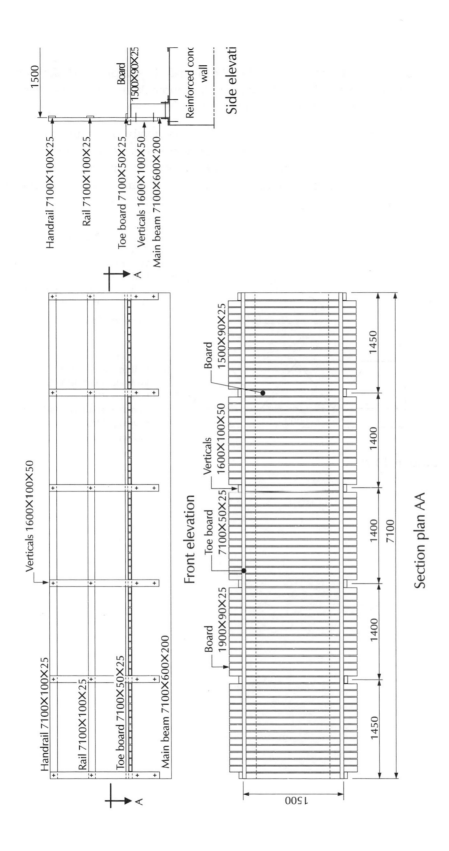

Handrail 7100X100X25

Rail 7100X100X25

Toe board 7100X50X25

Verticals 1600X100X50

Main beam 7100X600X200

Board
1500X90X25

Reinforced conc
wall

Side elevati

A

Front elevation

Board
1500X90X25

Verticals
1600X100X50

Toe board
7100X50X25

Board
1900X90X25

1450

1400

1400

1400

1450

7100

Section plan AA

1500

Figure 13.26 Detailing of a timber footbridge

13.6 Exercises

1 For the trusses described in Fig. 13.27 draw layout plan and elevation views showing the positions of trusses, purlins and columns. Also draw large scale detailed views of the truss joints.

2 Draw layout and detailed views of a timber footbridge 8 m span and 1.8 m wide. Choose one of the systems presented in Fig. 13.25a. Choose the sections of the main girders, handrails and decking boards as 200 mm × 700 mm, 25 mm × 75 mm and 25 mm × 100 mm respectively. Assume any missing data.

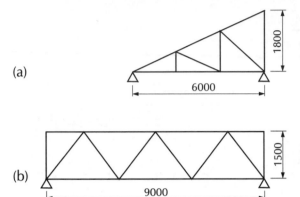

(a)

6000

Top chord 200×400
Bottom chord 200×400
Web members 200×300
Purlins 100×200
Concrete columns 300×300
Spacing of trusses 3 m
Number of trusses 7

1800

(b)

9000

Top chord 300×600
Bottom chord 300×600
Web members 300×500
Purlins 150×300
Concrete columns 300×500
Spacing of trusses 4 m
Number of trusses 6

1500

Figure 13.27

CHAPTER 14

Masonry Construction

14.1 Introduction

The term masonry construction is quite broad, covering a considerable range of materials and forms of construction. It has deep roots in many cultures reaching back to ancient times. Masonry is usually associated with a sense of permanency. Its durability, fire resistance, solidity and structural potential make it a logical and popular choice in many situations.

Structural masonry is presently used primarily for walls, piers and arches. In the past, it was also used for foundations, abutments and domes. Today, steel and reinforced concrete have replaced masonry, except perhaps in wall construction.

In this chapter, the main characters of masonry construction are introduced. The detailing rules of structures built of masonry are also explained with the help of practical examples.

14.2 Walls

General terminology

Figure 14.1 includes examples of masonry walls, piers, columns and pedestals. Piers are defined as parts of walls, with short lengths, while pedestals are very short columns.

Walls are usually laid up in horizontal rows, called courses, and vertical planes, called leaves. Most brick walls have only one or two leaves. If leaves are

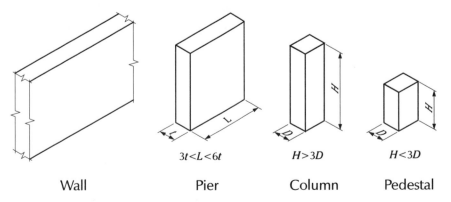

| Wall | Pier | Column | Pedestal |

$3t < L < 6t$ $H > 3D$ $H < 3D$

Figure 14.1 Masonry wall, pier, column and pedestal

connected directly, the construction is said to be solid, and if a space is left in between, the wall is called a cavity wall. If the cavity is filled with concrete, it is called a grouted cavity wall, and if reinforcement steel bars are inserted in the grouted cavity, the wall is called a reinforced wall (see Fig. 14.2).

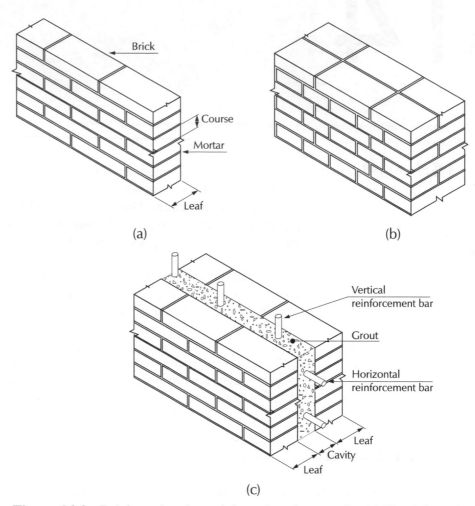

(a)   (b)

(c)

Figure 14.2   Reinforced and unreinforced masonry walls. (a) Unreinforced wall with one leaf. (b) Unreinforced wall with two leaves (solid construction). (c) Reinforced wall

Natural rocks and stones cut to shape are a well-known form of masonry used in building walls. Other forms include sun-dried, fired-clay and precast concrete bricks. Form, colour and structural properties vary considerably from one type to another, and even within one type. However, for all types of brick, the traditional size proportions of $L \times L/2 \times L/3$ are usually adopted. Cement mortar is commonly used as a binder.

Cavity walls   Cavity walls may be reinforced or unreinforced according to the structural use of the wall. In Britain, external cavity walls of short buildings (e.g. houses) are used for environmental reasons such as heat and sound insulation and hence are unreinforced. The bonding between the leaves of cavity walls (whether reinforced or unreinforced) is achieved using metal ties, some of which are shown in Fig. 14.3.

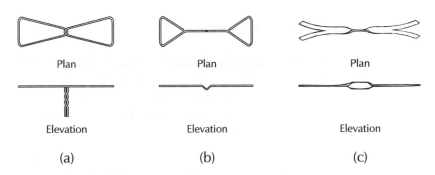

Figure 14.3 Metal ties used to tie leaves of cavity walls: (a) butterfly tie, (b) double triangle tie, (c) vertical twist tie

Brick positions and wall appearance

The ordinary position of a brick in a wall is when it is laid on its large side, with the long, thin side (face side) exposed. In such a case, the brick is called a stretcher (see Fig. 14.4). However, when two leaves are to be bonded together using masonry units, bricks may be turned to expose their ends, while laying large-side down or face-side down (see Figs 14.4a and 14.4b). A brick in these two positions is called a header and a rowlock respectively. For architectural purposes, a brick may be placed end-side down in the soldier position, shown in Fig. 14.4c.

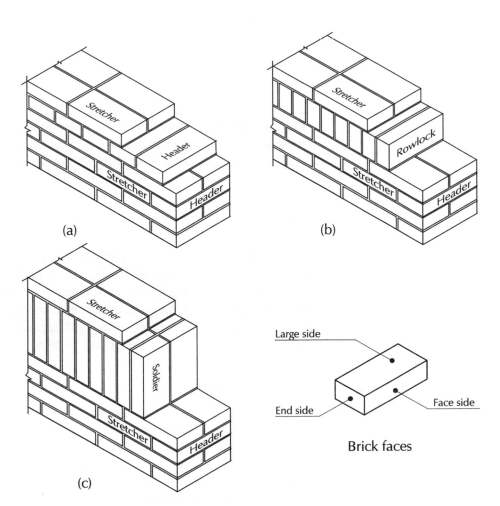

Figure 14.4 Possible brick positions in a brick wall

WALL FACE PATTERNS

The wall face patterns commonly used in practice are shown in Fig. 14.5. The stack pattern shown in Fig. 14.5f, which involves bricks in vertical and horizontal lines, is not sound if binding the leaves is fully dependent on the mortar. However, if the wall is fully grouted and reinforced, the stack pattern may be structurally acceptable.

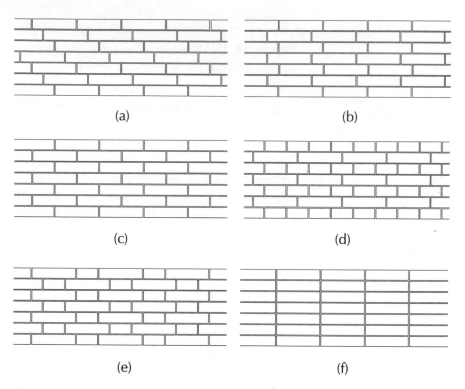

Figure 14.5   Face patterns of brick walls: (a) stretcher raking bond pattern, (b) stretcher quarter bond pattern, (c) common form of stretcher bond pattern, (d) English bond pattern, (e) Flemish bond pattern, (f) stack bond pattern

PROFILES OF MORTAR JOINTS

Mortar joints in a brick wall may have any of several profiles, the more common of which are the flush, weather struck, bucket handle and raked profile shown in Fig. 14.6. Bearing in mind that mortar joints make up about 17 to 24 per cent of the wall surface area depending on the bond pattern, it is clear that the joint profile has a significant effect on the wall appearance. Other important factors are mortar colour and texture, bond pattern, brick colour and the building quality.

Reinforcement and form variation

Masonry walls may be reinforced by incorporating steel bars in two directions embedded in the cavity between leaves. The steel bars used in reinforced masonry walls are identical to those used for reinforced concrete. The bars are commonly placed vertically and horizontally at intervals, and encased in concrete poured into the wall cavity. Masonry walls that do not have steel reinforcement are classified as unreinforced. Figure 14.2 shows typical reinforced and unreinforced masonry walls.

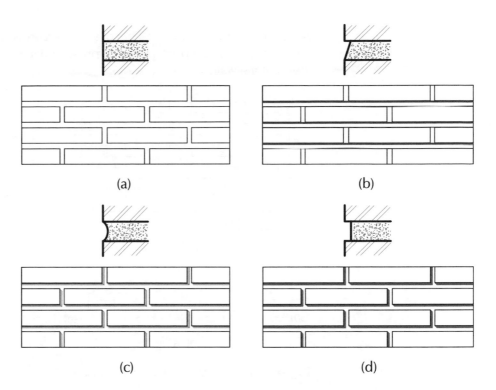

Figure 14.6 Mortar joint profiles: (a) flush joint, (b) weather struck joint, (c) bucket handle joint, (d) raked joint

Walls may also be strengthened by the use of form variation, examples of which are shown in Fig. 14.7. Turning a corner at the end of a wall (Fig. 14.7a), providing an enlargement in the form of a pilaster (Fig. 14.7b) and curving a wall in plan (Fig. 14.7c) add stability and strength to the wall. Building openings in a wall for doors or windows also requires a wall enhancement. Reinforced concrete beams (lintels) are often used to reinforce masonry walls at openings (see Fig. 14.7d). Alternatively, a masonry arch may be built as in Fig. 14.7e.

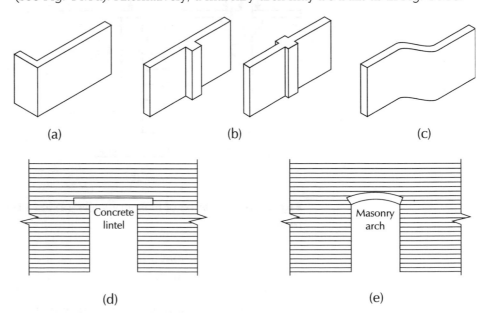

Figure 14.7 Methods to strengthen masonry walls: (a) sharp turn at end of wall, (b) pilaster projecting as one or both faces of wall, (c) wall curved in plan, (d) wall opening reinforced by a concrete lintel, (e) wall opening reinforced by a masonry arch

A typical example on the detailing of an unreinforced masonry wall with pilasters and a stretcher bond brick pattern is presented in Fig. 14.8 with three orthogonal views of the wall.

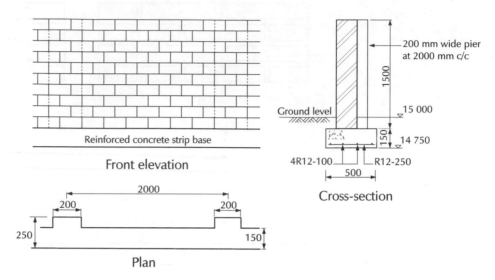

Figure 14.8 Masonry wall with pilasters

Use of walls in structures
Masonry walls are used for several purposes including:

■ Load-bearing walls carrying vertical weights of floors and roofs;
■ Exterior walls carrying wind pressures plus the common vertical loads of load-bearing walls;
■ Earth-retaining walls formed as cantilevers (see Fig. 14.9).

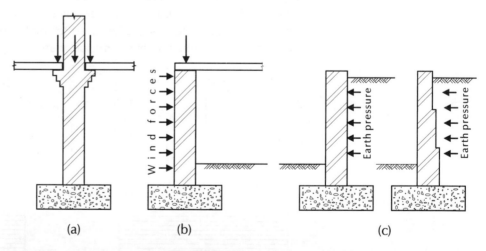

Figure 14.9 Types of masonry walls: (a) load-bearing, (b) exterior, (c) earth-retaining walls

Load-bearing walls may carry beams or slabs of floors in different ways, as shown in Fig. 14.10. Supporting devices attached to the face of the wall, such as the steel angle and the brick cantilever shown in Figs 14.10a and b, may be incorporated. Alternatively, a groove may be formed inside the wall for the floor to fit in. An example on the detailing of a reinforced masonry load-bearing wall is shown in Fig. 14.11a.

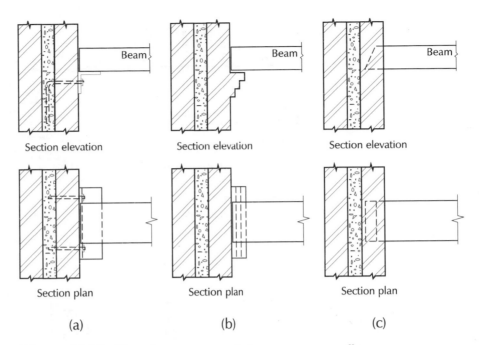

Figure 14.10 Floor beams supported on masonry walls

Reinforced masonry retaining walls are an attractive option in earth-retaining projects. In these walls, the two leaves must be tied together adequately and the wall reinforcement bars extended and bonded into the concrete base of the wall as shown in Fig. 14.11b. Notice that in Fig. 14.11 wall ties and reinforcement bars are detailed. A cross-section of the wall and base at a large scale is usually sufficient to show all wall details.

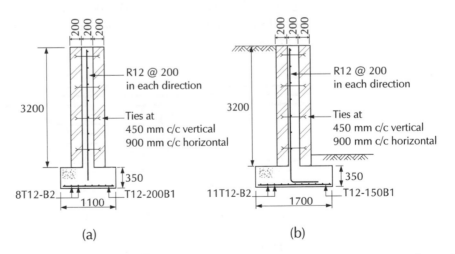

Figure 14.11 Detailing of masonry walls: (a) load-bearing, (b) earth-retaining walls

Example As a typical example of masonry structures, a two-storey house with masonry walls, reinforced concrete floor slabs and a pitched timber roof is discussed. A small scale section elevation of the building showing the main structural elements and joints is given in Fig. 14.12. A section plan would also typically be taken at each floor level and of bases to show the dimensions that are not given in elevation. The main joints of the building are also detailed (at a large scale) in Fig. 14.12. Notice the use of wall ties in external cavity walls. Internal walls would typically be composed of one leaf unless under heavy loadings.

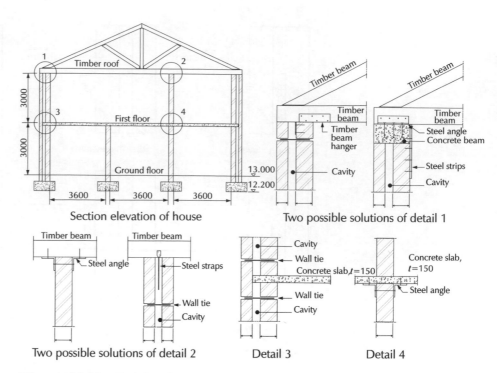

Figure 14.12   Details of masonry house

14.3 Columns

The smallest size of brick columns consists of only two bricks per layer, as shown in Fig. 14.13a. There is no opportunity to insert any vertical steel bars in this case, and the resulting column is unreinforced. Larger columns of side width equal to the length of a brick and a half, two bricks, or more, are popular since they can be grouted and reinforced (see Figs 14.13b and c). Columns may be used to receive vertical axial loads or to brace a wall like the pilaster shown in Fig. 14.7.

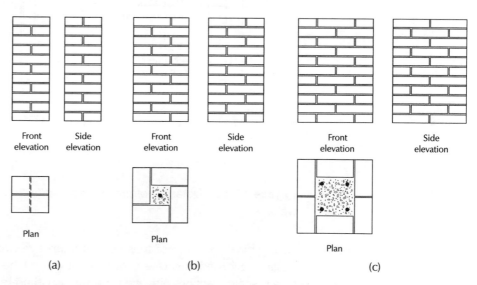

Figure 14.13   Masonry columns: (a) column with two bricks per layer, unreinforced, (b) column with four bricks per layer, reinforced, (c) column with six bricks per layer, reinforced

14.4 Masonry arches

Using masonry in arch construction is certainly one of the earliest innovative structural developments. The masonry arch saw major use in the spanning of wall openings for windows, doors, corridors and arcades. Today, it is still in use, but most probably is limited to decorative purposes. Figures 14.14 and 14.15 show some of the famous masonry arch patterns.

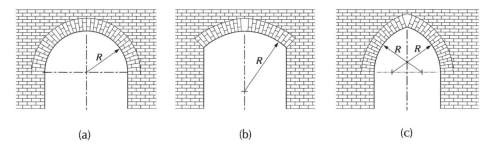

(a)   (b)   (c)

Figure 14.14   Masonry arches: (a) circular, (b) rowlock, (c) gothic arches

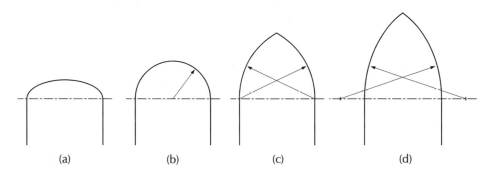

(a)   (b)   (c)   (d)

Figure 14.15   Further patterns of masonry arches: (a) elliptical, (b) semicircular, (c) equilateral gothic, (d) lancet

14.5 Stone masonry

For centuries, stone was a major construction material, used in building walls, columns, arches, domes, etc. Today, it is not a competitive structural material due to its high cost. The use of stone has now been reduced to building exteriors, mainly for decorative purposes.

Stone walls must be built with the same care as required for brick construction, in order to achieve a structurally sound form of construction. Among the patterns of stone masonry commonly used in practice are the random rubble, coursed rubble, random ashlar and coursed ashlar patterns (see Fig. 14.16).

Similar to brick walls, stone walls that carry major loads should be prepared with an internal cavity and therefore could be grouted and reinforced (see Fig. 14.17).

14.6 Exercises

1   Draw detailed orthogonal views for brick columns with cross-section sizes of $b \times b$, $1\frac{1}{2}b \times 1\frac{1}{2}b$, $2b \times 2b$ and $2\frac{1}{2}b \times 2\frac{1}{2}b$, where b is the length of a brick. The columns are grouted and reinforced if the size of the internal space allows.

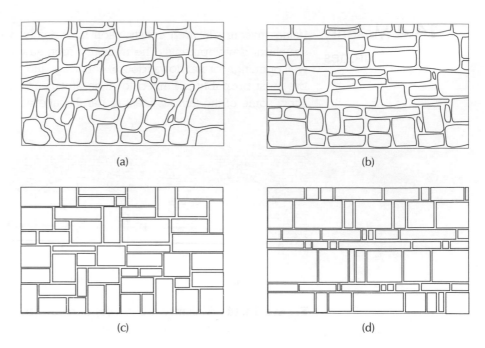

Figure 14.16   Face patterns of stone walls: (a) random rubble, (b) coursed rubble, (c) random ashlar and (d) coursed ashlar patterns

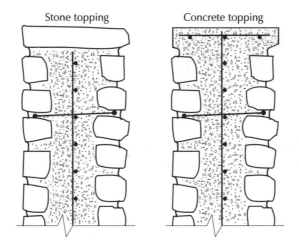

Figure 14.17   Reinforced stone walls

2   Draw large scale section elevations of the walls described below.
   (a)   An unreinforced load-bearing wall with one leaf, 0.2 m thick and 2.8 m high. The wall is built on a plain concrete strip footing of 0.8 m width and 0.3 m thickness.
   (b)   A reinforced load-bearing wall with two leaves, each of 0.2 m thickness, with a 0.25 m cavity in between. The wall height is 3.8 m and is reinforced with T12@250 mm in each direction. The wall is supported by a strip concrete base, 1.05 m wide and 0.35 m thick and reinforced with T12-200B1 and 8T12-150B2.
   (c)   A reinforced earth-retaining wall of the same dimensions and reinforcement as the wall in (b). The wall strip base is 1.65 m wide and 0.4 m thick and reinforced with T12-150B1 and 12T12-150B2.
   (d)   A reinforced load-bearing stone masonry wall with two leaves, each of 0.25 m average thickness. It has a 0.25 m internal cavity. The wall is

reinforced with T12@200 mm in each direction and is built on a concrete strip footing similar to that of wall (b).

3 Draw sketches of different ways to strengthen a masonry wall.

4 Draw detailed views of the wall shown in Fig. 14.18. The wall is 4.5 mm high and is made of reinforced masonry with concrete columns spaced at 5 m. The column reinforcement is 10T16 and the wall has T12@200 mm in each direction. The wall and columns are supported on a concrete strip footing, 1.9 m wide and 0.5 m thick, and reinforced with T16-150B1 and 14T12-150B2.

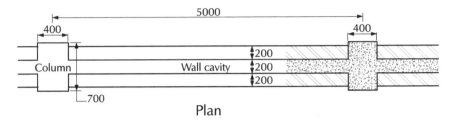

Plan

Figure 14.18

References and Further Reading

Baird, J. and E. Ozelton, *Timber Designers' Manual*, Granada Publishing Ltd, London, 1984.

Bangash, M., *Structural Details in Concrete*, Blackwell Scientific Publications, 1992.

Barnes, A. and A. Tilbrook, *The Theory and Practice of Drawing in SI Units*, The English Universities Press Ltd, London, 1970.

British Standards Institute, BS1192: Construction drawing practice.
Part 1. Recommendations for general principles, 1984.
Part 2. Recommendations for architectural and engineering drawings, 1987.

British Standards Institute, BS4466: Bending dimensions and scheduling of reinforcement for concrete, 1981.

British Standards Institute, BS8110: Structural use of concrete.
Part 1. Code of practice for design and construction, 1985.

British Standards Institute, PP7320: Construction drawing practice for schools, 1988.

Curtin, W., G. Shaw and J. Beck, *Structural Masonry Designers' Manual*, BSP Professional Books, London, 1991.

Curtin, W., G. Shaw, J. Beck and G. Parkinson, *Structural Masonry Detailing*, Granada Publishing Ltd, London, 1984.

Earle, J., *Drafting Technology*, second edition, Addison-Wesley Publishing Company, Wokingham, England, 1986.

French, T., C. Svensen, J. Helsel and B. Urbanick, *Mechanical Drawing*, ninth edition, McGraw-Hill Book Company, London, 1980.

Hayward, A. and F. Weare, *Steel Detailers' Manual*, Blackwell Scientific Publications, London, 1992.

Hoelscher, R., C. Springer and J. Dobrovolny, *Graphics for Engineers, Visualization, Communication and Design*, John Wiley and Sons, London, 1969.

Institution of Structural Engineers, The Concrete Society, *Standard Method of Detailing Structural Concrete*, 1989.

Karlsen, G., *Wooden Structures*, MIR Publishers, Moscow, 1967.

Knowlton, K., R. Beauchemin and P. Quinn, *Technical Freehand Drawing and Sketching*, McGraw-Hill Book Company, London, 1977.

MacGinley, T., *Structural Steelwork Calculations and Detailing*, Butterworths, London, 1990.

McKelvey, K., *Drawing for the Structural Concrete Engineer*, Viewpoint Publication, 1974.

Newton, P., *Structural Detailing: for Architecture, Building and Civil Engineering*, second edition, Macmillan Education Ltd, London, 1991.

Orton, A., *Structural Design of Masonry*, second edition, Longman, London, 1992.

Reiach Hall Blyth Partnership, *Footbridges in the Countryside, Design and Construction*, Countryside Commission for Scotland, 1981.

Reynolds, C. and J. Steedman, *Examples of the Design of Reinforced Concrete Buildings to BS8110*, fourth edition, E & FN Spon Publishers, 1992.

Shabaan, A., *Illustrated Metallic Bridges*, Alexandria University Press, Egypt, 1983.

Steel Construction Institute, *A Check List for Designers, Dimensions and Properties of Structural Steel Sections*, !986.

Timber Research and Development Association, Design examples to BS5268: Part 2: 1984.

Whittle, R. and T. Robin, *Reinforcement Detailing Manual*, Viewpoint publication, 1981.

Index